SOCIAL NETWORKS AND TRAVEL BEHAVIOUR

Social Networks and Travel Behaviour

Edited by

MATTHIAS KOWALD
Swiss Federal Office for Spatial Development, Berne

KAY W. AXHAUSEN
Swiss Federal Institute of Technology, Zurich

Routledge
Taylor & Francis Group
LONDON AND NEW YORK

First published 2015 by Ashgate Publishing

2 Park Square, Milton Park, Abingdon, Oxfordshire OX14 4RN
52 Vanderbilt Avenue, New York, NY 10017

Routledge is an imprint of the Taylor & Francis Group, an informa business

First issued in paperback 2020

British Library Cataloguing in Publication Data
A catalogue record for this book is available from the British Library

The Library of Congress has cataloged the printed edition as follows:
Social networks and travel behaviour / edited by Matthias Kowald and Kay W. Axhausen.
pages cm -- (Transport and society)
Includes bibliographical references and index.
ISBN 978-1-4724-3383-1 (hardback) -- 1. Travel--Social aspects. 2. Travel--Computer network resources. 3. Travel--Research. 4. Transportation demand management. 5. Transportation--Planning. I. Kowald, Matthias, editor. II. Axhausen, K. W., 1958- editor. III. Series: Transport and society.
G149.7S64 2015
306.4'819--dc23

2014039911

ISBN 978-1-4724-3383-1 (hbk)
ISBN 978-0-367-66883-9 (pbk)

Contents

Contents

List of Figures

List of Tables

List of Contributors

Theo Arentze

Dr Theo Arentze is Associated Professor of Urban Planning at the Eindhoven University of Technology. His research interests include activity-based modelling, discrete choice modelling, agent-based modelling, supernetwork modelling, human cognition/learning, and traveller information systems for application in transportation and urban planning. He is involved as principle researcher, supervisor or project leader in a constant stream of PhD, Postdoc and EU projects on these topics. He is a member of the editorial board of several international peer-reviewed journals and acts as an ad-hoc reviewer and program committee member for many journals, conferences and research foundations in transportation, planning, geography and consumer research.

Kay W. Axhausen

Dr K.W. Axhausen is Professor of Transport Planning at the Eidgenössische Technische Hochschule (ETH) Zürich (Swiss Federal Institute of Technology). He holds his post in the Institute for Transport Planning and Systems of the Department of Civil, Environmental and Geomatic Engineering. Before his appointment at ETH he worked at the Leopold-Franzens Universität, Innsbruck, Imperial College London and the University of Oxford. He holds a PhD in Civil Engineering from the Universität Karlsruhe (now KIT) and an MSc from the University of Wisconsin – Madison. He has been involved in the measurement and modelling of travel behaviour for the past 30 years contributing especially to the literature on stated preferences, micro-simulation of travel behaviour, valuation of travel time and its components, parking behaviour, activity scheduling and travel diary data collection. One strand of his current work focuses on the micro-simulation of daily travel behaviour and long-term mobility choices (see www.matsim.org for details). This work is supported by analyses of human activity spaces and their dependence on the traveller's personal social network. Recent work is experimenting with direct demand models as a radical alternative to disaggregate and aggregate models.

Juan-Antonio Carrasco

Juan-Antonio Carrasco is an Associate Professor at the Department of Civil Engineering, Universidad de Concepción. He holds a PhD in Transportation Engineering and Planning from University of Toronto, and a M.Sc. in Transport Engineering from P. Universidad Católica de Chile. His main research interest is on understanding and modelling the social dimension of travel behaviour, and the relevance of transport on equity and social exclusion. Currently, he is principal researcher at the Centre for Sustainable Urban Development (CEDEUS), research initiative between P. Universidad Católica de Chile and Universidad de Concepción, which seeks to have an integrated research perspective on mobility, heritage, planning, and environmental resources in Latin American cities.

Andreas Frei

Dr Andreas Frei is a Post-Doctoral Researcher and Adjunct Lecturer at the Transportation Center at Northwestern University. He received his doctoral degree from the Swiss Federal Institute of Technology in civil engineering. His research and interests in transport planning are in understanding and modelling of travel behaviour, especially the implications of social networks on travellers' choices and collective decision making. His dissertation work analysed and modelled trip making decisions beyond the usual measurable indicators, by extending the traveller's decision with a social network dimension.

In his career as a researcher, Dr Frei worked on several, mainly data driven projects in logistics, travel demand and supply analysis, optimization of systems and development of software tools. These projects were funded by the private industry (Echo Global Logistics, U.S. Bank Inc.) as well as the public sector, such as the Swiss National Science Foundation, the European Commission, the Federal Highway Administration and the Chicago Metropolitan Agency for Planning.

Matthias Kowald

Dr Matthias Kowald studied social sciences at University Duisburg-Essen where he graduated in 2007. From 2008 to 2012 he worked as a research assistant at the Institute of Transport Planning and Systems (IVT) of ETH Zürich. During this time his research addressed the influence of actors' social contacts on the individual travel behaviour. To particularly investigate social contacts' influence on leisure travel he managed a survey project that collected data on personal social networks and a population-wide network structure by taking a snowball sample. His research is summarized in his PhD-thesis 'Focusing on leisure travel: The link between spatial mobility, leisure acquaintances and social interactions'. Since 2013 he has worked at the Swiss Federal Office for Spatial Development (ARE) where he is in charge of national surveys on travel behaviour and related data analyses.

Timo Ohnmacht

Dr Timo Ohnmacht is a Professor in the Department of Economics at Lucerne University of Applied Sciences and Arts. He researches and teaches in the field of space, transport and society in the Competence Centre for Mobility. He studied Transportation and Sociology at the Technical University of Berlin, University of Lancaster (UK), and at the Swiss Federal Institute of Technology Zurich (ETHZ). One of his main activities is handling traffic statistics and analysing changes over time.

Fariya Sharmeen
Fariya Sharmeen is a PhD candidate at Eindhoven University of Technology. Her research interest is social contexts of urban and transportation planning and urban regeneration. Her PhD dissertation was focused on the relationship between social network dynamics and travel behaviour. She is a Commonwealth scholar and the recipient of the Royal Geographic Society Postgraduate award, 2013 for her work on the role of geography and path dependence on social interaction dynamics. Previously she worked as a Lecturer at Bangladesh University of Engineering and Technology, Bangladesh and as researcher at Cities Institute, UK and Utrecht University, the Netherlands. She has published several scholarly articles in renowned journals. She is as a member of the Education Committee of the Netherlands Graduate School of Urban and Regional Research and is a reviewer of leading journals such as Transportation Research Part A.

Harry J.P. Timmermans
Harry Timmermans is a Professor of Urban Planning at the Eindhoven University of Technology. His main research interests concern the study of human judgment and choice processes, mathematical modelling of urban systems and spatial interaction and choice patterns and the development of decision support and expert systems for application in urban planning. He has published several books and many articles in journals in the fields of Marketing, Urban Planning, Architecture and Urban Design, Geography, Environmental Psychology, Transportation Research, Urban and Regional Economics, Urban Sociology, Leisure Sciences and Computer Science.

Pauline von den Berg
Dr P.E.W. van den Berg is Assistant Professor of real estate management and urban development at Eindhoven University of Technology. She holds an MSc in Cultural Anthropology from Catholic University of Nijmegen, an MSc in Architecture, Building and Planning from Eindhoven University of Technology and a PhD in Urban Planning from Eindhoven University of Technology. Her PhD research focused on the role of social networks and ICT in social activity-travel patterns. Her research interests are in the areas of the built environment and quality of life, personal social network analysis, the spatial distribution of leisure and social activities and their consequences for urban and transport planning. Her work has been published in leading journals (e.g. Transportation, Transportation Research Part A, C, Transportation Research Record, Environment and Planning A, B, Journal of Transport Geography, Journal of Geographical Systems, etc.).

Introduction

Transport Planning and Social Network Analysis – An Introduction

Matthias Kowald and Kay W. Axhausen

Transport planning began to use the methods of social network analysis about 15 years ago (see for example, Axhausen (2005) based on a presentation given in 2002; also see Chapter 1). Aiming to understand and model peoples' out-of-home movements and related choices like traffic mode and location choice, the academic field of transport planning was confronted with two major challenges at that point: First, various national travel surveys in Western societies indicated an increasing importance of the leisure travel segment (for examples, see BFS and ARE 2012; Deutsches Institut für Wirtschaftsforschung 2012; DfT 2012; U.S. Department of Transportation 2011). However, most work in transport planning focused on travel related to work purposes, i.e. commuting and peak hour traffic causing many problems for transport system managers (Larsen et al., 2006). Approaching and explaining leisure travel, as an example of non-work related traffic, is in contrast more complex. In comparison to work related behavioural patterns, peoples' variety in leisure behaviour is usually much higher: For example, leisure travel and activities are less rigid in temporal and spatial patterns and they are more influenced by external factors like social contacts or weather conditions (see Schlich et al., 2004).

The second challenge resulted from the development and availability of disaggregated modelling frameworks to approach (personal) travel. Traditional aggregated models are zone based. In these model frameworks, trips between zones are determined by the attractiveness of a destination zone to perform certain activities and the generalized costs between origin and destination. The value of an activity is the utility a person gains when satisfying a need. Needs are developed in the zone of origin and can be satisfied at one or several destination(s). Need satisfaction naturally accounts for generalized costs like travel time and monetary costs that usually increase with an increasing distance between two zones. The choice to perform a certain activity at a certain place is therefore explained by a person's characteristics including individual constraints in time, money, and capability, and the generalized costs. Decision makers are assumed as being independent from each other, exclusively driven by their individual needs. Therefore, these traditional aggregated models do not include any information on peoples' social interactions or their personal social networks. In contrast, disaggregated models are often implemented in an activity based way. This allows a more detailed modelling of travel including traditional elements like individual characteristics, motivations, constraints and travel costs, and, in addition, a consideration of new influences from an actor's social environment as well. Considering peoples' interactions means that a person does not exclusively travel within an infrastructure but also within a social structure (Frei, 2012). Therefore, decision makers are not considered as being independent from but as influencing each other.

These two main driving factors made transport planners focus on peoples' interaction and their social network. In recent years there have been a remarkable number of data collection efforts in the field, surveying information on the link between travel behaviour and social motivation. Providing an overview on selected exemplary studies is the main purpose of this volume. The overview includes three dimensions: First, the motivation of the overlap between transport planning and the methods of social network analysis is addressed, including a detailed discussion of the research questions. Second, an overview of the applied methods of social network analysis and related empirical results is provided. Third, current challenges and new research questions in this area are outlined.

Focusing on a coherent sequence of research studies in Switzerland helps to trace back the use of social network analysis methods in transport planning. It can be seen how first hypotheses emerge from theoretical considerations and literature research (Chapter 1). These first hypotheses led to a qualitative study including in-depth interviews to gather first insights on the spatial spread of respondents' social contacts and related mobility strategies to maintain these relationships (Chapter 2). The experience of this first study was the basis of a larger-scale quantitative survey study on peoples' personal networks (Chapter 3). This study particularly focused on respondents' network size, the spatial spread of their social contacts and communication strategies to maintain these relations. It thus aims to test hypotheses and results from Chapters 1 and 2 empirically. However, usually personal networks are embedded in a population-wide network structure. This underlying network allows feelings (Christakis and Fowler, 2009), styles and behaviours (Gladwell, 2002), as well as contagious diseases (Mossong et al., 2008; Smieszek et al., 2011) to spread within a given population. It can be assumed that the level of social coherence and related network issues also influence travel behaviour decisions. Accordingly, a second survey study aimed to reproduce the (local) findings on personal networks from Chapter 3 in a nation-wide survey and study the population-wide network structure of the Swiss population (Chapter 4). Chapters 2, 3 and 4 include detailed information on the survey strategy, instrument and data analysis. They thus show in which way the methods of social network analyses can be applied in transport planning.

Three additional chapters introduce current challenges and research questions. Chapter 5 focuses on differences between personal networks from different populations and locations. A comparison between empirical results from various survey studies, including Swiss and non-Swiss data collection efforts, identifies such differences. The analysis particularly focuses on influences among socio-demographics, personal network characteristics, mobility biographical aspects, social network geographies and spatial patterns between social contacts by considering the local, regional, or national context of each study and the time they were in the field (Chapter 5). Chapter 6 compares issues of personal networks from Toronto, Canada and Concepción, Chile. This second comparison study focuses on effects from travel demand and supply to enable and maintain social relationships by considering the available transport systems. Finally, Chapter 7 highlights the time dynamics of social networks. As people establish new relationships and maintain or loose older ones, their personal network structures change over time. An overview of the time patterns of personal networks is accompanied by a discussion on the importance of this dimension for transport planning issues.

This edited volume on the link between social contacts and travel behaviour patterns fits well into the series Transport and Society. This becomes clear when recalling an earlier volume of this series, *Mobilities, Networks, Geographies* by Larsen, Urry and Axhausen (2006) or the volume edited by Grieco and Urry (2012) *Mobilities: New Perspectives on Transport and Society*. Building on these, the current volume presents exemplary research studies. It is by no means the last word on the use of social network analysis methods in transport planning but provides an intermediate overview on the methods applied and the insights obtained. Chapters can be read consecutively and include cross-references wherever necessary. However, it is also possible to focus on selected chapters as each of them introduces its research question, technical terms.

References

Axhausen, K.W. (2005) Social networks and travel: Some hypotheses, in K. Donaghy, S. Poppelreuter and G. Rudinger (eds), *Social Dimensions of Sustainable Transport: Transatlantic Perspectives*, Aldershot: Ashgate, pp. 90–108.

BFS and ARE (2012) Mobilität in der Schweiz, Ergebnisse des Mikrozensus Mobilität und Verkehr 2010, Federal Office for Spatial Development and Swiss Federal Statistical Office, Bern and Neuenburg.

Christakis, N.A. and J.H. Fowler (2009) *Connected: The Surprising Power of our Social Networks and How They Shape Our Lives*, New York: Little Brown.

Deutsches Institut für Wirtschaftsforschung (2012) Mobilität in Deutschland 2008, www.mobilitaet-in-deutschland.de/pdf/MiD2008_Tabellenband.pdf, accessed December, 2012.

DfT (2012) Purpose share – average distance travelled: Great Britain, https://www.gov.uk/government/uploads/system/uploads/attachment_data/file/35607/nts0402.xls, accessed December, 2012.

Frei, A. (2012) Networks, geographies and travel: Travel between infrastructure and social structure, Dissertation, ETH Zurich, Zurich.

Gladwell, M. (2002) *The Tipping Point: How Little Things Can Make a Big Difference*, Boston: Back Bay Books.

Schlich, R., Schönfelder, S. Hanson, S. and Axhausen, K.W. (2004) Structures of leisure travel: Temporal and spatial variability, *Transport Reviews*, 24(2): 219–37.

Larsen, J., Urry, J. and Axhausen, K.W. (2006) *Mobilities, Networks, Geographies*, Aldershot: Ashgate.

Mossong, J., N. Hens, M. Jit, P. Beutels, K. Auranen et al. (2008) Social contacts and mixing patterns relevant to the spread of infectious diseases, *PLoS Med*, 5(3): 381–91.

Smieszek, T., Balmer, M., Hattendorf, J., Axhausen, K.W., Zinsstag, J. and Scholz, R.W. (2011) Reconstructing the 2003/2004 H3N2 influenza epidemic in Switzerland with a spatially explicit, individual-based model, http://www.ncbi.nlm.nih.gov/pmc/articles/PMC3112096/, accessed January 2013.

U.S. Department of Transportation (2011) 2009 National Household Travel Survey. Summary of Travel Trends: Average Annual PMT, Person Trips and Trip Length by Trip Purpose, http://www.google.ch/url?sa=t&rct=j&q=&esrc=s&source=web&cd=1&ved=0CDMQFjAA&url=http%3A%2F%2Fnhts.ornl.gov%2F2009%2Fpub%2Fstt.pdf&ei=_VflUJzCD4eSswaFp4DgAQ&usg=AFQjCNHzYpw2DY0xAe2jpEHXwvTFpm9MxA&sig2=VtEOUnDc_q7K43MaXF9MHg&bvm=bv.1355534169,d.Yms, accessed December, 2012.

Chapter 1
Activity Spaces, Biographies, Social Networks and their Welfare Gains and Externalities: Some Hypotheses and Empirical Results[1]

Kay W. Axhausen

The literatures on urban sprawl (see e.g. Axhausen 2000 for a review), or globalization (e.g. Friedman 2005 or Taylor 2004) take today's historically low generalized costs of travel either as given, or see it as something surprising. Neither attitude is helpful, if one wants to understand the social impact of this seminal change and the related changes in travel behaviour. The investment in transport infrastructures and in the rolling stock exploiting their possibilities is a necessary part of the modernization and industrialization of first the North-Atlantic societies and today the world (see Axhausen 2005 for the argument). The purpose of this chapter is not macro-sociological, but micro-sociological and will therefore not explore the links between market size, productivity growth and real incomes or between generalized cost of transport, food availability and population growth (see Sen 1983). It wants to explore a set of linked hypotheses about activity space as the result of the structure and geography of a traveller's social network. It should be pointed out that these hypotheses should be taken as the starting points for yet to be undertaken empirical work. The evidence, which I can adduce here, is based on very small samples, incomplete, often from questions smuggled into other surveys.

The accelerated shrinking of distance through infrastructure additions allowing ever increasing sustained speeds has now been going on for two hundred years (see Grübler 1992 and 1998 for the period since 1800, but Spufford 2002 or de Vries and van der Woude 1997 for earlier periods). Canals, pike roads, then the inter-urban and urban railways were the main investments during the nineteenth century. These infrastructures lowered transport costs already very considerably, but for many travellers and much daily traffic this early motorization remained irrelevant, as the vehicle costs remained too high relative to many incomes. The nickel-fares of late nineteenth-century US local public transport, and then the

1 This chapter is a reprint of the paper, which appeared in 2007 under the same title in *Mobilities*, 2(1): 15–36. The references have been updated where necessary.

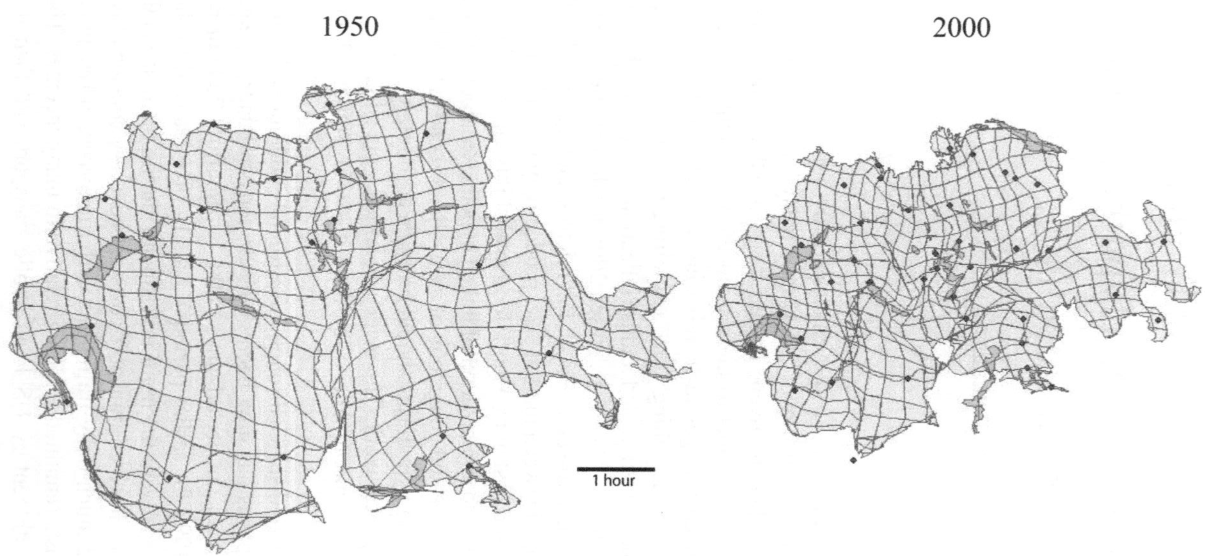

Figure 1.1 Road travel time – scaled mapping of Switzerland (same scale)

Source: Maps: Scherer (2004) and later revisions.

cheap and reliable car, liberated the urban travellers from their dependence on their feet. Figure 1.1 illustrates this for post-war Switzerland using time-scaled maps (see Scherer 2004 for the creation of the maps; Fröhlich and Axhausen 2005 for network models employed). The maps, in which distances indicate travel time, show that the country has essentially been more than halved in size, which is even more important when one considers that the motorization rate went from 31 for 1000 inhabitants in 1950 to 492 in 2000 (Widmer and Meister 2005). While the costs for the usage of cars have stagnated during the last decades (Abay 2000 for Switzerland and Rhode and Strumpf 2003 for the USA), the quality-adjusted purchase costs continue to decline, as shown in Figure 1.2. The even more dramatic drop in the costs of telecommunication (Figure 1.3) is in line with the developments in physical transport. Zero marginal cost local calls are becoming the rule worldwide; zero marginal cost long-distance calls are becoming available through voice-over-IP technology (see for example www.skype.com).

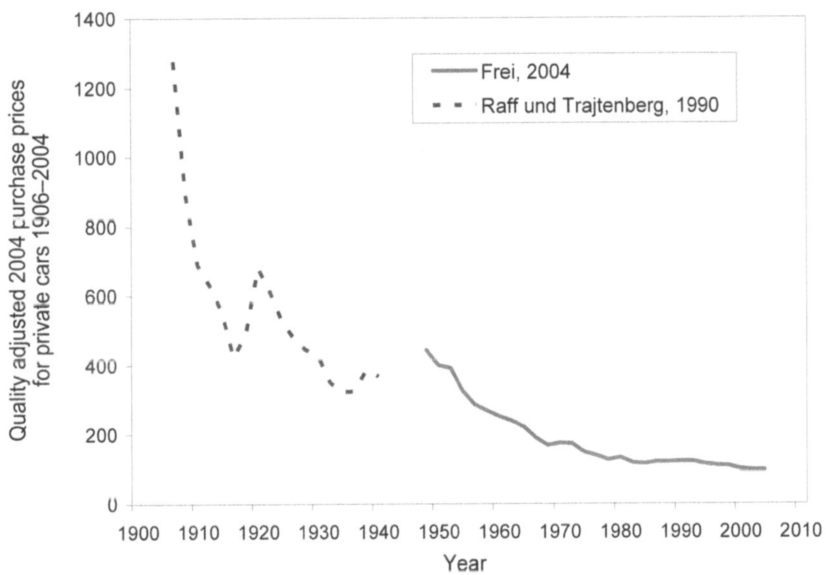

Figure 1.2 Quality adjusted 2004 purchase prices for private cars 1906–2004

Source: Data: Frei (2005) for Switzerland 1950–2004; Raff and Trajtenberg (1985) for the USA, 1906–1940. The Swiss time series is very likely underestimating the drop in the quality adjusted prices since 1985, as appropriate time series data on vehicle electronics were not available for analysis.

The impact of this small world or better cheap world in generalized cost terms, on the structure of the social networks of the travellers and their daily life will be discussed in the following section. The 'small world' issue discussed since Milgram's initial paper (1967) in for example Travers and Milgram 1969; Watts 1999; Barabasi 2002 or Dodds, Muhamad and Watts 2002 will be left aside as an essentially academic exercise.[2] Structure means here not only the topology of the network (Wasserman and Faust 1994), but also its geography. The topology describes in a link/node notation which persons in an environment are linked to the others. A unidirectional link exists if a certain minimum flow of resources can be observed over a defined period from one person to a certain other person. The corresponding reverse link exists normally, but not necessarily, as well. The resources can be classified according to type, as for example money, labour, attention, social recognition, emotional support, time for a particular joint activity. It is assumed that the volume of flow across types, but also the breadth of the types involved is a measure of the strengths of the link.

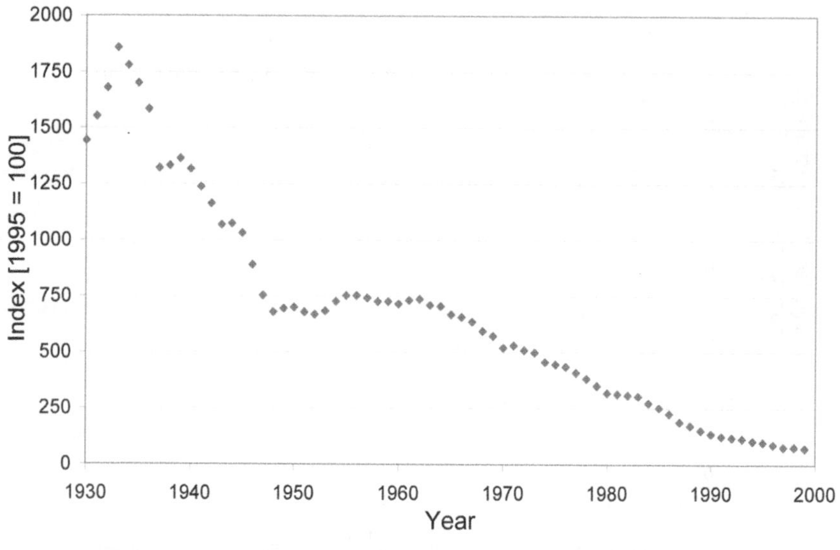

Figure 1.3 Real costs of US interstate and international telephone calls 1930–2000

Source: Adapted from FCC 2001.

2 The enormous attrition rates of the 'small world' experiment reported in Dodds, Muhamad and Watts, 2002 raise the question, if and if so, how quickly information will spread in such networks (see Huberman and Adamic 2005 for a review of this issue).

The relative strengths of the flows between a pair of persons can be interpreted as a measure of their relative power. The analysis of the management of these links in the pursuit of daily life, medium-term projects and life goals is a core subject of micro-sociological enquiry (see for example Goffman 1959; Fehr and Fischbacher 2004 or Fehr and Gächter 2000; Grieco 1996 for discussions of such central topics as management of self, fairness and reciprocity, local trust and household resource management).

The social science literature on social networks (e.g. Wellman 1996, 1999a, b, 2000, 2001; Moore and Brown 1970; Hampton and Wellman 2001, 2003; Haythornwaite and Wellman 1998; Freeman 2000; Burt 2000b and 2001) tends to ignore or aggregate space, so that the information about the relative locations of the actors and therefore their travel and communication costs is lost. This geography of the relative locations and the associated costs of travel and communication is the focus of the interest here. The assumption is that effective changes in the geography of costs and locations will impact both travel and social patterns profoundly. The geography of the members of the traveller's social networks together with the geography of the relevant activity locations determines the amount and style of travel. Just consider that most leisure travel, which now makes up the relative majority of all trips in Switzerland for example, involves travelling with and meeting others. See Figure 1.4 for mean number of household and non-household members travelling with the 250 respondents over a 6-week reporting period (Axhausen, Löchl, Schlich, Buhl and Widmer 2007 describe the survey in detail). This implies that the locations will reflect the joint preferences of the group, potentially living at dispersed locations.

For the individual any system-wide change runs in parallel with his/her personal development. His/her mobility biography of intertwined changes in home, education and work locations and availability, ownership of mobility tools (licences, cycles, motorcycles, cars, season tickets and public transport discount cars) and social networks will reflect those changes, but also modulate it according to personal needs (for an example emphasizing network geographies see Blinde and Schlich 2002). In the interpretation of empirical results one should keep this in mind.

This double understanding of structure creates a problem in the use of the words: dense and density, which can refer to both the spatial density (number of objects per area), as well as to the density of linkage in a social network (number of links originating from a person per person). To avoid confusion, this chapter will use the opposites 'loose, tight' to describe the patterns of linkage, while the pair 'sparse, dense' will refer to the spatial density.

In principle one cannot divide the total network linking humans at any one time: still, in everyday practice persons will treat their total network, as if it divides into separate networks based on their common activities, their genesis, their common awareness of each other, their permissible questions and demands, their solidarities.

The chapter will be structured as follows. In the next section, it will discuss the problem of how to measure the size and spread of the social network geographies and of daily life in an efficient and parsimonious way, so that change or difference can be captured. The main section will propose and discuss the set of hypotheses. Finally the chapter will summarize the hypotheses by embedding them in a simple model of personal resource allocation and by offering a research programme.

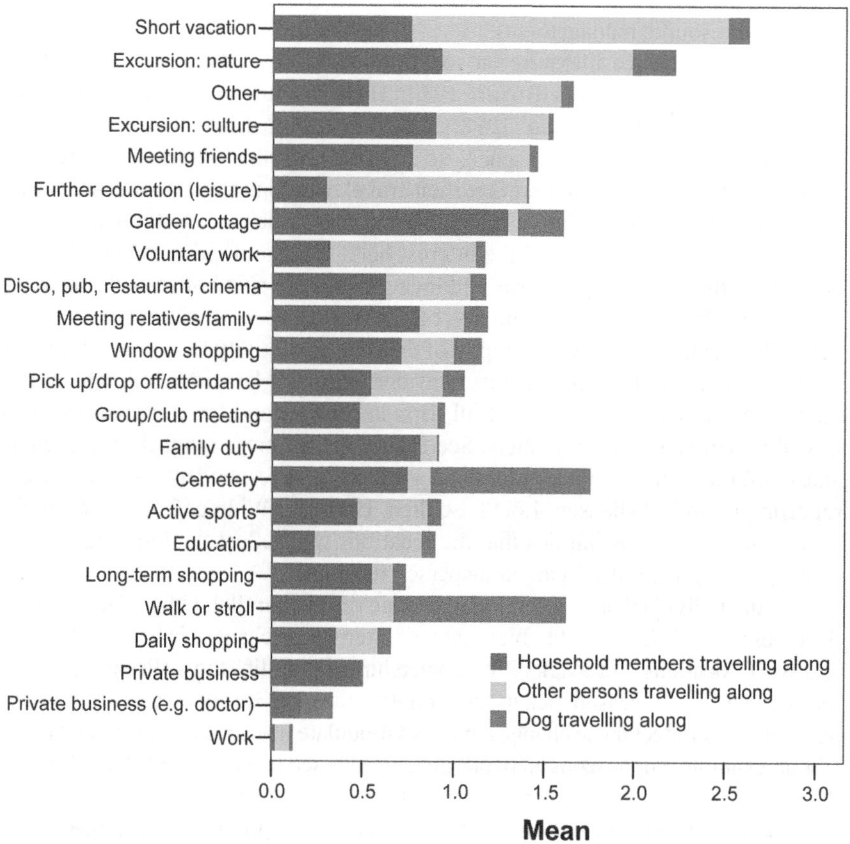

Figure 1.4 Mean number of persons travelling with the respondents of the 2003 Thurgau six-week travel diary survey by trip purpose

Measuring the Network Geographies and Activity Spaces

The main static elements of the geography of a social network are the home locations of its membership, while the dynamic elements of its activity space are the locations where some or all members meet. If the network is defined through joint association with an organization, such as a firm, or club, then the seat of this organization could be counted among the static elements of the geography. Given the range of networks and the increasing fluidity of their static elements, such as the workplace, it seems preferable not do so. Maps showing the locations of the homes or joint activities are very useful, but while the analyst can detect patterns quickly these patterns need to be reduced to a small set of indicators, which can compared between networks, traced over time and modelled. Schönfelder and Axhausen 2003a, b, suggest a series of such measures for individual activity spaces. The confidence ellipse, the two-dimensional generalization of the confidence interval, is the easiest to compute of the proposed measures.[3] As a parametric approach (fixed geometrical form), it will overestimate the size of the geography, in particular, if one chooses a point other than the weighted mean of the locations as its centre. Still, they show that it is strongly correlated with more complex approaches, which seem therefore to be redundant. Figure 1.5 shows the home locations of the significant contacts of a person interviewed by Ohnmacht (2004), as part of a study on travel, biographies and social networks in Zürich (see Chapter 2). It also shows the 95% confidence ellipse around the mean home location over the life-course of the respondent. The ellipse captures both the spread and the orientation of the network geography. The ellipse seems therefore an appropriate measure for network work geographies in the first instance; especially given its ease of computation.

Figure 1.5 Example of a geography of the friendship network

Source: Ohnmacht 2004 (Person 4; also see Chapter 2); 95% confidence ellipse is calculated around the mean x and y coordinates of the home locations of the respondent over her life course.

---

3 The confidence interval is that part of the real numbers, which will include the true, population mean value with a specified confidence and its calculated using the mean and variance of the sample at hand. For normally distributed values it will also contain 95% of the values of the population.

The usefulness of this approach for the tracing of change is visible in Figure 1.6. Here the confidence ellipse is used to describe the in-commuter sheds of the ten largest Swiss cities for each decade since 1970. Their growth over the last three decades, but also their change in orientation becomes clearly visible.

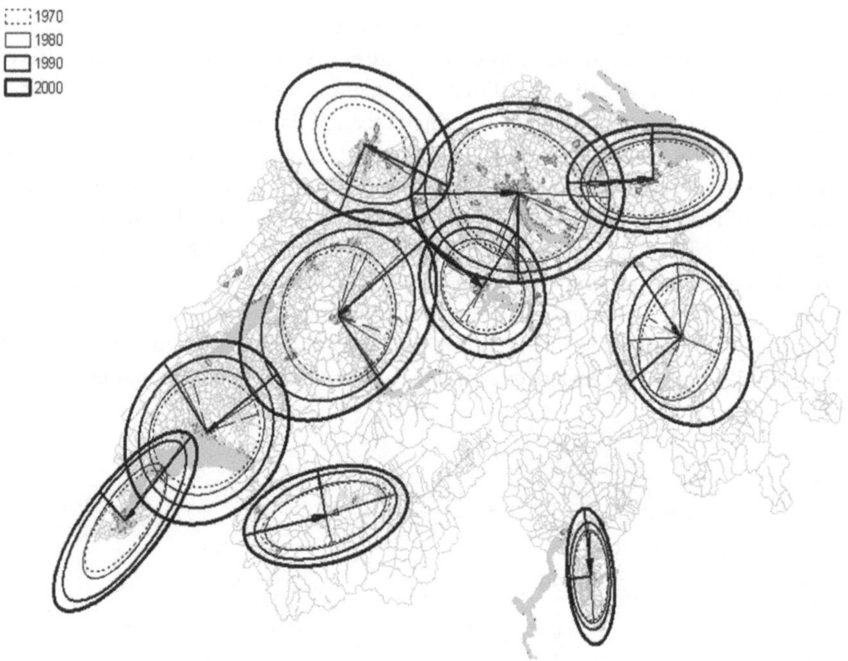

Figure 1.6 In-commuter-shed of the 10 largest Swiss cities since 1970
Source: Adapted from Botte 2003.

These two examples demonstrate that one can make useful statements about the size of a social network geography, which are based on the spread of the relevant locations. This is important, as the even simpler measures based on the distributions of the distances between the individual and its contacts, such as the mean or median, do not contain information about the spread of the locations.

Hypotheses

The hypotheses and their corollaries will address in turn the size of network geographies, the number of contacts and networks, their structure and finally the expected impacts on travel behaviour and welfare.

The first hypothesis is:

[1] The size of the social network geography is inversely proportional to the generalized costs of travel and communication. Individuals use decreases in generalized costs to meet and retain contacts living further away. It is well known that travellers adjust their travel distances in response to system improvements. See for example Cervero and Hansen 2000 or Fröhlich 2003 for evidence from models of induced demand or Schäfer and Victor 2000; Zahavi and Talvitie 1980 on the constant travel time budget hypothesis. These increased distances imply also social contacts at increasing distances given the importance of visiting friends, family and colleagues among all travel, which is borne out by the fact that the share of leisure and business trips increases with increasing trip distance.

A corollary of this hypothesis is that

[1.1] social contacts should become more selective, as persons can choose among a larger number of possible contacts available within the same generalized cost isoline, i.e. the homophily of most social networks. There is no need any more to socialize with spatial neighbours. This is consistent with the generally observed trend to individualization (Giddens 1984 or Beck 1986), which implies more specialized patterns of exchange and social matching. A further corollary is that

[1.2] the mobile phone increases social selectivity by making it more difficult to meet persons without prior coordination, as persons are less likely to be reliably in certain places at certain times.

A third corollary is, that

[1.3] the spatial density of contacts falls with decreasing generalized costs of travel, as the result of the first two suggestions.

The second hypothesis is:

[2] The number of contacts individuals maintain is inversely proportional to the generalized costs of travel and communication. As defined above, a social network link only exists if there is a certain volume of exchange between the two sides. As the basic exchange of information, gossip and therefore attention becomes more affordable, one would expect that individuals do not drop contacts, but maintain them using the available telecommunication channels, but also using the possibilities to multiply messages nearly freely employing for example the cc line of the email system, a professional mailing list, or Xerox copies of newsletters. See also Burt 2000a.

On the other side it is unlikely that the ability to co-ordinate a network grows at the same rate as the exponentially growing number of potential interactions within a growing network, one would therefore expect as a corollary, that

[2.1] Individuals will belong to more networks as the number of active contacts grows. This is consistent with Dunbar's (1992) suggestion that membership of social groups (networks) have upper limits, as they splinter when the time needed for the *social grooming* essential for its coherence becomes prohibitive. It is also consistent with the group size scaling laws proposed in Zhou, Sornette, Hill and Dunbar (2005).

The distribution of the increasing number of contacts in an increasing number of networks can be accommodated by different topologies, in particular with different link densities (counting here links between the same persons separately, if they arise from different contexts and networks, such as work, church, sport club, card playing round). Given the implied co-ordination effort, one would expect with the third hypothesis:

[3] The probability of being linked to a member of one's network through multiple networks increases with the spatial density of one's contacts. This drop in the spatial social (network) density should reduce the casual social supervision of the local environment, as it contains fewer known and therefore significant others. Jane Jacob's (1961) 'eyes on street' disappear for lack of interest. See Watts and Strogatz (1998), Newman (2001) or Girvan and Newman (2002) for empirical measurement of such clustering in non-localized professional contexts.

Even with longer life spans and lower average working hours the 24 hour time constraint remains in the short-run and therefore as a limitation on individuals in the management of their social networks through the allocation of time and other resources; implicitly time for those dependent on their wages. While the absolute size of the number of social contacts has received substantial attention (see example Pool and Kochen 1978; Bernard, Johnsen, Killworth and Robinson 1987; Grierson, Smith and Wain 2003; Van Tilburg 1995; McCarty, Bernard, Killworth, Shelley and Johnsen 1997 or Zheng, Salganikz and Gelman 2005), the distribution of the effort among these contacts seems to have received less attention. From a transport perspective this information is crucial, as it is correlated with the likelihood of travel. While time budget studies could be designed to provide such information, they presently do not, as they neither identify individuals contacted, visited or spoken to nor cover enough days for a reliable estimate. Indirect measures, such as visiting frequency by type of contact, provide some insight, but they omit detailed information about other forms of contact. The person survey of 2003 Thurgau diary asked the respondents to nominate up to 5 relatives and up to 5 other important contacts and to specify both the meeting frequency and their home location. Using the crow-fly distance between the home locations of the respondents and their contacts one can estimate the effort by approximating it with the distance travelled (Figure 1.7) and the distribution of total effort, here for the sample as a whole (Figure 1.8). See Licoppe and Smoreda (2005) for the amount of phoning before and after a move or Axhausen, Frei and Ohnmacht (2006) for the market shares of face-to-face meetings, phone, SMS and email contact as a function of distance (also see Chapter 3).

It is likely that not only the distribution of this crude proxy is left skewed, but also the true distribution of effort among the social contacts, in particular if one considers, that the household members are missing in this data set. Hypothesis four is:

[4] The distribution of effort on non-household members will become more left skewed as the spatial social network density decreases. A socially tightly knit local environment provides a large amount of information about contacts and contacts of contacts. It affords the opportunity to meet others without scheduling the meeting, as one knows where one is likely to meet another person, quite

independently of random meetings, which will also occur more frequently when using the same small and shared environment. This density of information means that the other often does not need to explain much to sketch a situation or problem. This information externality is lost when the local social tightness is reduced and requires contacts to spend more time on updating the other about everyday details, which form the background of any larger issue. Together with the stronger selectivity mentioned above, it is likely that most persons will reallocate time and resources to maintain a certain quality of interaction with their most important contacts leading to the stipulated shift. The lower costs involved in maintaining contacts with the lower ranked contacts supports this shift as well.

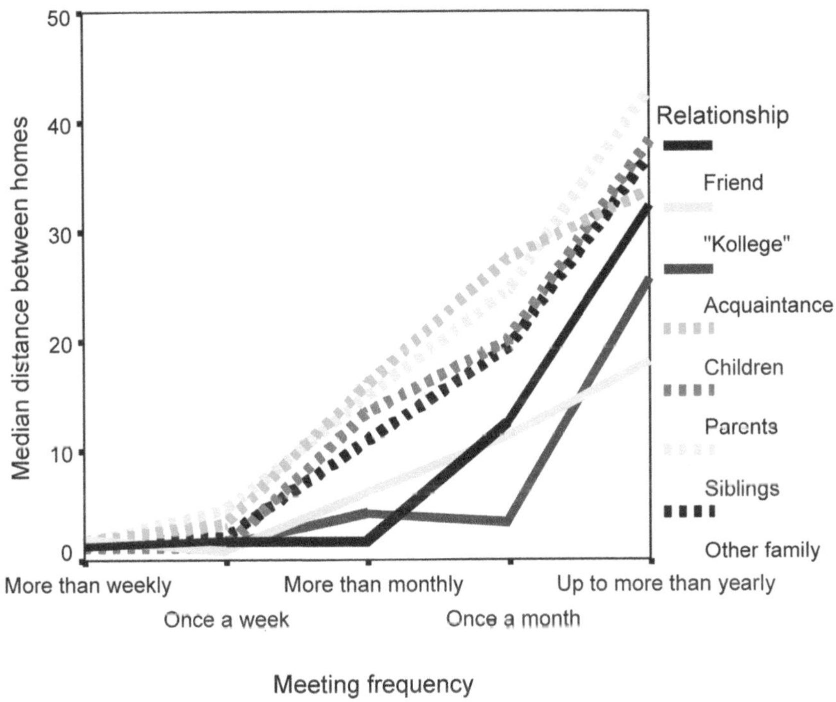

Figure 1.7 Median distance between home locations by type of relationship and frequency of visit (2003 Thurgau survey)

Note: 'Kollege' in Swiss-German is a term covering everybody from workmate, schoolmate, fellow member of a sports clubs, fellow soldier to somebody who is more than just an acquaintance. Note as well, that it is well known, that people overestimate the frequency of rare events.

In a spatially dense and therefore strongly overlapping set of social networks it is likely that members know about the contacts of their contacts, even if these are not part of their own social network. The stipulated shift in the distribution of effort and in the local density leads therefore to hypothesis five:

[5] The knowledge about the contacts of contacts in a social network is proportional to the generalized costs of travel and communication. This change, if true, has the externality, that the efficiency of the network to spread information, gossip or rumour locally is reduced. It has also the corollary, that

[5.1] The distance decay of the social supervision is less steep in spatially looser sets of networks. If a set of social networks is spatially well defined, then its members know when they have left its territory, left the range of supervision. The 'dirty weekend in Brighton' reflected that distance in Victorian Britain. If sets of networks are spatially less well defined, and more spread out, the chance increases to meet a friend of a friend, of whom one had not been aware, further away from home. This should be true not only in the spatial sense, but also in the sense of social distance.

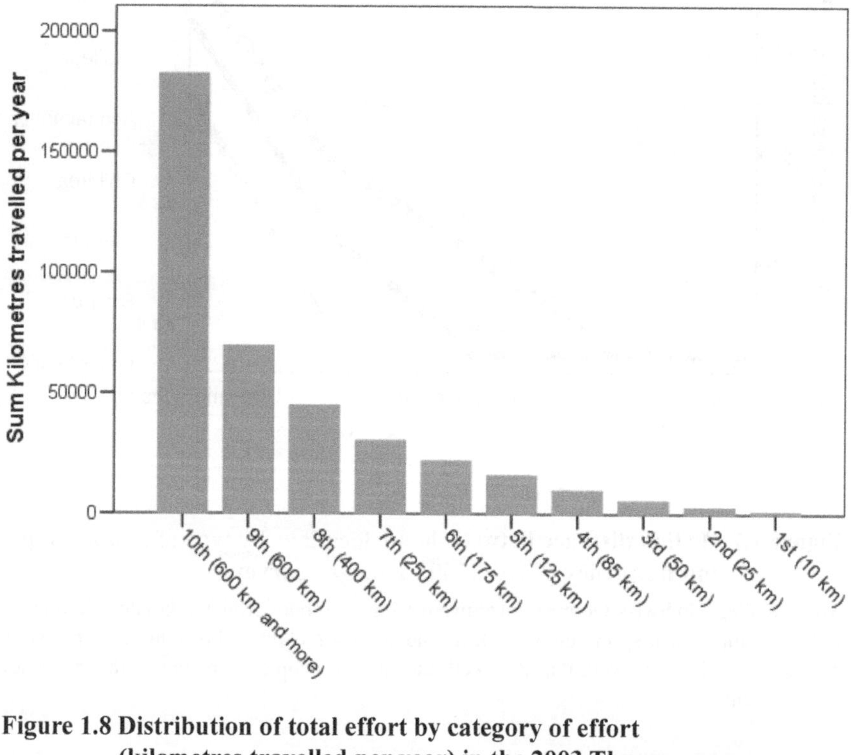

Figure 1.8 Distribution of total effort by category of effort (kilometres travelled per year) in the 2003 Thurgau survey

Note: The effort distribution was divided in deciles; the labels are approximate.

Social milieus reinforce this effect. Social milieu can be understood as the voluntary recreation of the small world of high generalized costs of travel through understood sets of meeting points and common events. The type, style and combinations of meeting points and common events differ between milieus; e.g. university professors of a discipline will have for example joint conferences, preferred institutions for sabbaticals, typical clients and derived from these locations for other professional and informal meetings (holiday locations around conference locations; restaurants around the offices of important clients, etc.). This is reinforced through the common preference for non-professional activities given an often shared habitus in Bourdieu's (1984) sense. As a person belonging to a milieu is likely to frequent these common meeting points anywhere it finds itself, the likelihood increases for the person that he/she will and can connect with friends of friends, or formulated negatively reports about behaviour away from home will reach home quickly. The mobile phone reinforces this through its recreation of a common locality through the web of SMS, chat and talk.

The mobile phone (Geser 2005) makes milieus (social networks) without fixed locations and events possible, as the coordination can be achieved on the move. The substantial and sophisticated coordination performances around recent *anti-globalization* events are good examples of these possibilities.

These *structures of encounterability*,[4] of who is likely to be where, when, with whom, are known or can be identified using established clues for different milieus, in particular if they have formed into defined subcultures, such as those around religious affiliation, political belief, diasporas or sexual orientation. Modern creations, such as alumni networks, fan clubs, etc., add to these structures of predictability in a world, which appears locally socially unstructured.

In addition, such structures are constructed for further subgroups of the population through their joint scheduling necessities or infrastructure use: think of parents of a particular type or aspiration congregating at the end of the school day at the relevant school; consider the commuters' time-space paths bundled into the same train and carriage through the railway timetable for a specific origin and destination.

If social network geographies and social network structures behave as stipulated in the hypotheses and corollaries discussed above, then this should have impacts on travel behaviour, the level of local trust and the welfare of the individuals. The increased geographies require more travel in terms of miles or kilometres travelled, as personal meetings remain essential to the coherence and development of a network for the time being (Urry 2000 and 2003). The distances involved for these meetings will sustain a demand for high-speed travel, which in turn can become more efficient exploiting economies of scale and scope; here larger airplanes and

4 Margaret Grieco suggested this concept during discussions in the course of the Dft-funded project 'Social networks and future mobilities: 'Meetings' in the twenty-first century'.

more routes/lines. Furthermore, the social network geography is linked to the activity space, i.e. the geography of daily travel, in the sixth hypothesis:

[6] The activity space of an individual is proportional to its social network geography. The action space, i.e. the geography of locations known to a person, will grow even faster, as the increasing spatial spread and increasing number of contacts will combine to raise awareness of yet not visited locations. This trend is overlaid with the aging of the individual, who is accumulating information over his/her lifetime. The attitudes towards travel, communication and openness towards others and other things (variety seeking) are formed during the childhood and youth of an individual. We would expect therefore correlations between the activity space and social network geographies between generations. This is an additional effect beyond the part of the social network an individual inherits from his or her parents.

Finally, while the action space continues to grow (Hypothesis seven),

[7a] The size of the local activity space of an individual stabilizes after an initial exploration.

[7b] The size of the total activity space will grow in line with the growth of social network geographies over longer periods. New locations will tend to be further away from the home location than previously visited locations, as likelihood of finding a new location is larger outside the existing range. Their small share of all locations shifts the footprint from month to month. Still, there should be an equilibrium size, when the rate of turnover and change inside the activity space produces enough new locations to satisfy the variety seeking of the individual. The size of the total activity space will grow in line with the growth of social network geographies over longer periods.

The reduction in the number of interested 'eyes on the street' and the reduction in the local tightness of network links should result in increased feelings of insecurity, independent of the actual number of crimes. The environment becomes increasingly socially empty and therefore less trustworthy (see Seligman 2000 for the problem of social trust). Localized anomie in a society, which is still well integrated at larger scales should become more frequent (Franzen 2000). (See Axhausen 2000 for a review of the recent critique of suburbanization, which is partially based on these fears.)

This contrast between the local and non-local comes to the fore in times of crisis, whether for serious cases such as illness or accidents, supervision of children or elderly relatives, but also for trivial cases, such as the care of potted plants or the emptying of a letterbox during a prolonged absence. Hypothesis eight is, that

[8] The reliance on commercial or publicly funded personal services increases proportionally with the geography of social networks as services traditionally traded or undertaken in local social networks cannot be secured anymore due to the lower densities of contacts. The commercialization of such services produces different patterns of stress for their users, as a) time gets traded against money, which needs to be earned (see Becker 1976), and b) as some form of supervision is still required due to the possible unreliability or incompetence of such services.

Finally, the social selectivity possible in a world of cheap travel and telecommunication should result in social more homogenous networks (see also the review of McPherson, Smith-Lovin and Cook 2001). These in turn should become more productive for their members (see Galor and Weil 2000 for the economic success inherent in specialization and increases in the economies of scale). The balance between social and psychological challenge and reinforcement can be set by each individual through the range of networks, to which it belongs. Those networks, which are not constructed according to such *Wahlverwandschaften* (elective affinities), i.e. family, and less so work, church, profession, should become less important and come under increasing stress. All told, hypothesis nine is, that on balance

[9] the welfare of the individuals should increase inversely proportional to the generalized costs of travel as they have more control over their social context, even accounting for the costs identified above. One question though is how the costs of the local anomie are distributed between different strata of the population. The world of low generalized travel costs still favours the wealthier, which can respond easier to unwelcome change through the exercise of their *exit option* (Hirshman 1970). The strength of the local organized *voice option* in the pursuit of the management, redirection or reversal of such change is likely to be weaker as a result. Equally, the number of weak ties in Granovetter's (1973) sense is likely to be reduced for persons who are relatively more bound to the locality. While much of the factual information formerly brokered by the weak ties can be found today in electronic forms, the essential qualitative assessment of the information and the sponsorship available from the weak ties is reduced.

A Qualitative Model of Social Networks and Activity Space

The hypotheses suggested above stand alone and are not linked to other choices an individual makes. Ignoring the biographical dimension, Figure 1.9 suggests a quasi-static model linking individual choices to the size of individual activity space and the structure and geography of the personal social networks. It indicates the sign of the elasticities (rule of proportionality), but not their size. Microeconomic theory suggests that the individual can increase his wages by specialization or by increasing the number of hours worked. Investment into the level of specialization through education and further training is deemed to be the better long term strategy, which is not only adopted by individual families, but also as national policy by most countries around the world. A condition for successful specialization is that the labour market has to be large enough to offer a relevant post or posts. From an individual's point of view specialisation requires more often than not either migration or long commutes, which in turn are easier if the generalized costs of travel and communication are small.

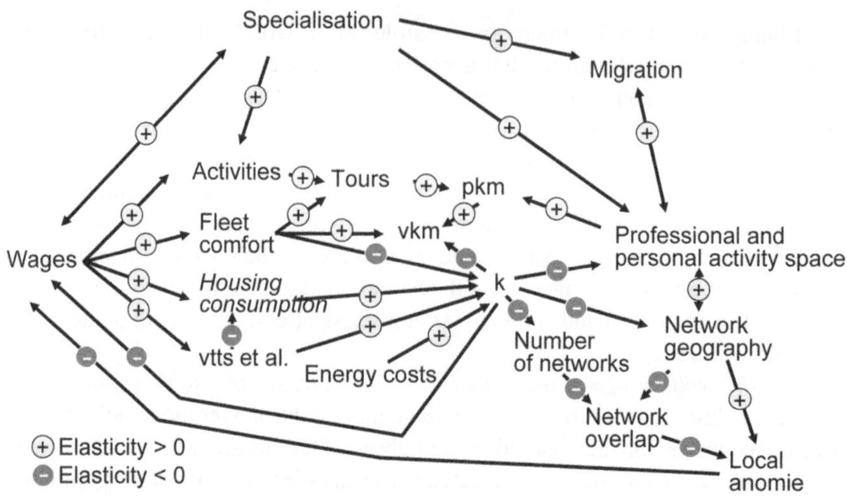

Figure 1.9 Expanded qualitative model of social networks and activity space

Note: k: Generalized costs of travel (and communication); pkm: Personal km; vkm: Vehicle km; vtts: Value of travel time savings; fleet comfort: number and standard of the personal mobility tools, i.e. cycles, cars, trucks and public transport season tickets; The circles indicate the sign of the assumed elasticity or proportionality (plus = positive, proportional; minus = negative, inversely proportional).

While all consumer goods are elastic with respect to income and wages, the model focuses on those important for travel behaviour. Increases in wages do encourage out of home activity, as does the division of labour inherent in specialization, even if these effects will be small with respect to the number of trips undertaken. The income elasticities of housing consumption and fleet comfort are known to be substantial (see for example Rumley 1984; Hanly, Dargay and Goodwin 2002; Fogel 2002; Barker 2003 or Barot and Yang 2002). The central point about the vehicle fleet or more generally the set of mobility tools (cycle, motorcycle, car, SUV, season ticket, public transport season tickets – transit passes) which a person acquires, is that lump sum payments are exchanged for lower variable costs. A private car allows the owner to avoid the lower speed and lower flexibility of public transport or the higher variable cost of a taxi ride. All public transport rides are free (at the point of use) for the season ticket holder.

The willingness to pay for lower travel times, better higher speeds, more comfort and more reliability are known to be elastic to income (see for example König, Abay and Axhausen 2004 or Fosgerau 2005), which increases the generalized costs of travel under *ceteris paribus* conditions. The same is generally also true for increased housing consumption, as this normally requires locations further from the denser parts of an urbanized area. Still, unless collective choices keep capacities or slots for movement very tight, one would expect that in sum the individual is able to lower the generalized costs of travel, which implicitly increases its wages (see Cervero and Hanson 2002 or Levinson and Karamalaputi 2003 for highway network supply elasticities).

The model in its lower right hand side then summarizes the hypotheses discussed above, while removing some detail for clarity. This detail is added in Figure 1.10.

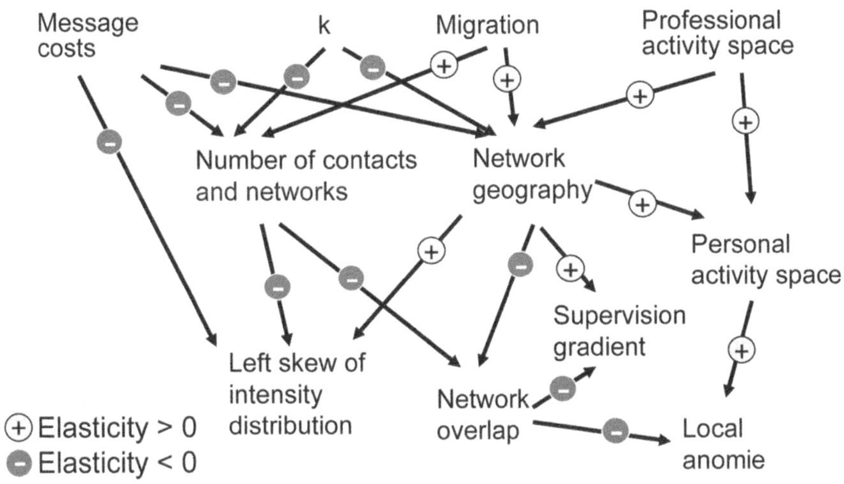

Figure 1.10 Hypotheses summarized

Note: k: Generalized costs of travel; The circles indicate the sign of the assumed elasticity or proportionality (plus = positive, proportional; minus = negative, inversely proportional).

It is interesting to note that the growth in vehicle kilometres driven is sustained through three mechanisms as long as real incomes rise: increased fleet comfort, increased housing consumption, lower generalized costs of travel and increased activity spaces, such as those shown in Figure 1.6. The monetary and social investment in these areas makes it likely that the process is not easy to reverse once under way. A hysteresis like pattern is likely to emerge due to the social capital inherent in the networks, such as common understandings, trained ability for successful joint performance, co-evolution of skills or joint history (see also Lin 2001; Sobel 2002; Grieco 1988). In such a pattern the path taken during the phase of falling generalized costs will not be the same as the path, if the costs should increase again.

Having assumed throughout, that the observed and stipulated processes go hand in hand with increased levels of personal welfare, the model also explains the on-going dissolution of the local social coherence typical for a world of high generalized costs of travel and communications. In that world individuals were linked to each other through multiple contexts and were likely to meet each other regularly – planned or by chance (Tönnies 1887 – reprint 1935). The last decades can only be described as a flight from this environment for most people (Putnam 1999), who seem on average to prefer a spatially less tight social environment. Still, some groups never wanted to leave, such as orthodox Jews tied to their synagogue or the Amish tied to their community by the refusal to use cars; other people want to recreate it in parts by buying into gated communities (Low 2003 or Blakely and Snyders 1997) or new urbanist environments, which help to identify others with the wish to live a more local life; some people were not able to leave, as they did not have and still do not have the resources to do so. Manifold problems arise for the last type of people, as they have to live in a small world, when the rest of the world assumes that it is a large world, which is driven by the processes alluded in the introduction: market growth through specialization, economies of scale and growing markets.

Research Questions

The qualitative models and the hypotheses formulated above form a research programme as many of the stipulated links and effects have not been empirically demonstrated yet. There are three areas of urgent work at this point: solving the measurement problems, estimating the behavioural elasticities and understanding the links between social welfare, localized anomie and the size of the activity spaces.

There are various approaches to obtain a complete or nearly complete view of the set of contacts of an individual. Diaries can be supplemented by personal telephone books, email directories, telephone bills, photo albums, yearbooks, rosters of the clubs and associations, internal phone books of employers etc. In addition, various tests based on name recognition can be constructed. While such work is challenging enough, the real challenge is the reconstruction of past patterns, as the persons need to remember where their contacts lived at a particular time.

Current surveys show that respondents in face-to-face interviews are willing and able to reconstruct their mobility biography and their network of important contacts over time (see Ohnmacht 2004 and ongoing work at Lancaster with respondent which were recruited using a substantial incentive payment (SFr75 (€50) or £20).[5] A recent paper-and-pencil survey involving a complex matrix representation of time and items (homes, workplaces, mobility tools) resulted in good quality responses, but a low response rate (Axhausen, Frei and Ohnmacht 2006; also see Chapter 3). In neither case have the answers of the respondents been validated independently using official records or third parties.

This issue of validation is even more important for retrospective answers, as few people will have retained their address books over the years. The switch

5 Since published as Larson, Urry and Axhausen (2006).

to electronic databases makes such personal record keeping even less likely in the future.

One central research area is therefore the development of reliable and cost-efficient methods to record and reconstruct the social network geographies. Validation of the respondents' answers through either official or written records or snowball sampling of the identified contacts will need to be explored.

The available methods for tracing the daily activity space, diaries, but in particular satellite positioning systems and mobile phone records are proven, even if in need of fine tuning. Here the issue is the reconstruction of past activity spaces from respondents' recollections. There is a need to test new approaches, as current efforts using a small number of supposedly key locations have not been successful so far.

The models and the hypotheses above suggest the behavioural system responds to changes in the structure of the generalized costs. Some of these elasticities are known (for some countries), but many are not, in particular those concerning the structure and geographies of the social networks and linked with them of the activity spaces. While these estimates would be interesting in their own right, they are especially important for transport policymaking. Much of the policy discussion in transport is focusing on increasing the generalized costs of travel in line with the external costs of travel and sometimes beyond. If the hypotheses presented here are correct, it is unlikely that the adjustment of the individuals will be fast due to the investment into the contacts behind the travel, nor politically easy to achieve.

The chapter has also suggested a further externality of the new social network geographies: localized anomie due to an under-critical local social network tightness. Transport researchers have addressed this issue obliquely through their interest in the feeling of personal safety while walking, especially to and from public transport stops. In the transport and regional planning literature studies have not linked their results to the local social network structures. The measurement problem is substantial as both the feeling of safety under different conditions and the local social network tightness have to be established. If the link can be established, then the question remains, if this externality should be addressed socially and if so how, or if the personal responses of exit (moving away) and individual protection (fortifying the residence and the use of motorized personal and protected spaces (cars)) are appropriate. See Brook and Durlauf 2001 or Durlauf and Cohen-Cole 2004 for the difficulties of model estimation, when social structure is endogenous to the issue at hand.

In this context one could make use of recent work on the general happiness of a population to see, if some of the variation of the level of happiness can be explained by the structure and geographies of the social networks (see for example Stutzer and Frey 2004 or Frey and Stutzer 2001). Making this link would allow us to see, if continuing private and social investment in lower generalized cost of travel do indeed increase the social welfare further.

Policy Issues and Ideas: A Brief Coda

This chapter is not about the policy. Still, a number of ideas and issues should be briefly restated and highlighted here. If the set of hypotheses outlined above are true, then one would have to rethink the policy imperative of improving accessibility, connectivity and (physical and social) mobility. This is seen as a panacea for the reduction of spatial inequalities, e.g. between peripheral and central regions of a country or Europe, for the reduction of spatially concentrated social exclusion, for the revitalization of economically depressed areas. These potential gains would have to be weighed against possible increases in localized anomie, the lack of trust in the immediate environment of other fully, but non-locally socially integrated populations. This externality would have to be added to the usual set of environmental and economic externalities arising from generally inappropriately priced infrastructure used. Here, the costs incurred by the population to protect themselves against perceived threats or to purchase services formally provided informally among neighbours, are actually more easily measurable then the environmental costs of traffic.

Acknowledgements

This chapter develops ideas, which have been presented and discussed on a number of occasions (Axhausen 2000, 2002, 2005, 2006). The author is grateful for the suggestions and critique, which arose at seminars organized by M. Grieco, Cornell University, Martin Lee-Gosslin, Laval University, Julian Hine, University of Ulster, Fritz Busch, Technische Universität München, Barbara Lenz, DLR, Berlin, Uwe Serkölt, Universität Zürich, and Ph. Toint, FUNDP, Namur. The on-going discussions with M. Grieco have shaped this chapter in many ways.

The chapter has also benefited from the joint project 'Social networks and future mobilities: "Meetings" in the twenty-first century' with J. Urry and J. Larson at the University of Lancaster funded by the Horizons Programme of the UK Department for Transport and with T. Ohnmacht, TU Berlin on 'Mobility biographies, mobility tools and social networks' funded by the Institut für Mobilitätsforschung, Berlin.

References

Abay, G. (2000) Preisentwicklung im Personenverkehr, report to the ARE, Abay und Meier, Zürich.

Axhausen, K.W. (2000) Geographies of somewhere: A review of urban literature, *Urban Studies* 37(10): 1849–64.

Axhausen, K.W. (2002) A dynamic understanding of travel demand: A sketch, *Arbeitsberichte Verkehrs- und Raumplanung*, 119, Institut für Verkehrsplanung und Transportsysteme, ETH Zürich, Zurich.

Axhausen, K.W. (2005) Social networks and travel: Some hypotheses, in K. Donaghy, S. Poppelreuter and G. Rudinger (eds) *Social Dimensions of Sustainable Transport: Transatlantic Perspectives*. Aldershot: Ashgate, pp. 90–108.

Axhausen, K.W. (2006) Social factors in future travel: A qualitative assessment, *IEE Proceedings Intelligent Transport Systems* 153(2): 156–66.

Axhausen, K.W., Frei, A. and Ohnmacht, T. (2006) Networks, biographies and travel: First empirical and methodological results. Paper presented at the *11th International Conference on Travel Behaviour Research*, Kyoto, August 2006.

Axhausen, K.W., Löchl, M., Schlich, R., Buhl, T. and Widmer, P. (Forthcoming) Fatigue in long duration surveys, *Transportation* 34(2): 143–60.

Barabasi, A.L. (2002) *Linked: How Everything Is Connected to Everything Else and What It Means*. New York: Plume.

Barker, K. (2003) *Securing our Future Housing Needs – Review of Housing Supply: Interim Report – Analysis*, report to HM Treasury, London: http://news.bbc.co.uk/1/shared/spl/hi/uk/03/budget/documents/pdf/barker_review_foretoch3_396.pdf.

Barot, B. and Yang, Z. (2002) House prices and housing investment in Sweden and the United Kingdom: Econometric analysis for the period 1970–1998, *Working Paper* 80. Stockholm: National Institute of Economic Research.

Beck, U. (1986) *Risikogesellschaft*. Frankfurt: Suhrkamp.

Becker, G.S. (1976) *The Economic Approach to Human Behavior*. Chicago: University of Chicago Press.

Bernard, H.R., Johnsen, E.C., Killworth, P.D. and Robinson, S. (1987) Estimating the size of an average personal network and of an event subpopulation: Some empirical results, *Proceedings*, American Statistical Organisation: http://www.amstat.org/sections/srms/Proceedings/papers/1987_003.pdf.

Blakely, E. and Snyders, M.G. (1997) *Fortress America*. Washington, DC: Brookings Institution.

Blinde, J. and Schlich, R. (2002) Freizeitmobilität und Wohnsituation, in M. Gather and A. Kagermeier (eds) *Freizeitverkehr, Hintergründe, Probleme Perspektiven, Studien zur Mobilitäts- und Verkehrsforschung*, 1. Mannheim: Verlag MetaGIS Infosysteme, pp. 35–53.

Botte, M. (2003) Strukturen des Pendelns in der Schweiz, MSc thesis, Fakultät für Bauingenieurwesen, TU Dresden, August 2003.

Bourdieu, P. (1984) *Distinction: A Social Critique of the Judgement of Taste*. Cambridge, MA: Harvard University Press.

Brock, W. and Durlauf, S. (2001) Discrete Choice with Social Interactions, *Review of Economic Studies* 68(2): 235–60.

Burt, R.S. (2000a) Decay Functions, *Social Networks* 22(1): 1–28.

Burt, R.S. (2000b) The Network Structure of Social Capital, in R.I. Sutton and B.M. Straw (eds) *Research in Organization Behaviour* vol. 22. Greenwich: JAI Press, pp. 345–423.

Burt, R.S. (2001) Structural holes versus network closure as social capital, in N. Lin, K. Cook and R.S. Burt (eds) *Social Capital: Theory and Research.* New York: Aldine de Gruyter, pp. 31–56.

Cervero, R. and Hansen, M. (2002) Induced travel demand and induced road investment: A simultaneous equation analysis, *Journal of Transport and Policy* 36(3): 469–90.

Dodds, P.S., Muhamad, R. and Watts, D.J. (2002) An experimental study of search in global social networks, *Science* 301: 827–29.

Dunbar, R.I.M (1992) Neocortex size as a constraint on group size in primates, *Journal of Human Evolution*, 20(xx): 469–93.

Durlauf, S. and Cohen-Cole, E. (2004) Social interaction models, *Working Paper*, Economics Department, University of Wisconsin, Madison: http://www.ssc.wisc.edu/econ/Durlauf/research.html.

FCC (2001) *Long Distance Telecommunication Industry*. Washington, DC: FCC.

Fehr, E. and Gächter, S. (2000) Fairness and retaliation: The economics of reciprocity, *Journal of Economic Perspectives* 14(1): 159–81.

Fehr, E. and Fischbacher, U. (2004) Social norms and human cooperation, *Trends in Cognitive Sciences* 8(4): 185–90.

Fogel, R.W. (2002) Toward a postmodern egalitarian agenda. Paper presented at the *Commonwealth North Forum*, Anchorage, August 2002.

Fosgerau. M. (2005) Unit income elasticity of the VTTS. Paper presented at the 8th Nectar Conference, Las Palmas, June 2005.

Franzen, A. (2000) Does the Internet make us lonely?, *European Sociological Review* 16(4): 427–38.

Freeman, L.C. (2000) Social network analysis: Definition and history, in A.E. Kazdan (ed.) *Encyclopedia of Psychology* vol. 6. New York: Oxford University Press, pp. 350–51.

Frei, A. (2005) Was hätte man 1960 für einen Sharan bezahlt?, MSc Thesis, Department Bau, Umwelt und Geomatik, ETH Zürich, Zürich.

Frey, B.S. and Stutzer, A. (2001) *Happiness and Economics: How the Economy and Institutions Affect Human Well-Being*. Princeton, NJ: Princeton University Press.

Friedman, T.L. (2005) *The World Is Flat: A Brief History of the Twenty-First Century*. New York: Farrar, Straus and Giroux.

Fröhlich, P. (2003) Induced traffic: Review of the explanatory models. Paper presented at the 3rd Swiss Transport Research Conference, Ascona, March 2003.

Galor, O. and Weil, D.N. (2000) Population, technology, and growth: From Malthusian stagnation to the demographic transition and beyond, *American Economic Review* 90(4): 806–28.

Geser, H. (2004) *Towards a Sociological Theory of the Mobile Phone*, Version 3. Zürich: Soziologisches Institut, Universität Zürich.

Giddens, A. (1984) *The Constitution of Society*. Cambridge: Polity Press.

Girvan, M. and Newman, M.E.J. (2002) Community structure in social and biological networks, *Proceedings of the National Academy of Sciences of the USA* 99(12): 7821–26.

Goffman, E. (1959) *The Presentation of Self in Everyday Life*. New York: Anchor Books.

Granovetter, M. (1973) The strength of weak ties, *American Journal of Sociology* 78(6): 1360–80.

Grieco, M.S. (1988) *Keeping It in the Family: Social Networks and Employment Chance*. London: Tavistock Publications.

Grieco, M.S. (1996) *Workers' Dilemmas: Recruitment, Reliability and Repeated Exchange*. London: Routledge.

Grierson, J., Smith, A. and Wain, D. (2003) *Victorian Networks Study: Technical Report Method, Sample and Preliminary Findings*. Melbourne: Latrobe University.

Grübler, A. (1990) *The Rise and Fall of Infrastructure: Dynamics of Evolution and Technological Change in Transport*. Heidelberg: Physica-Verlag.

Grübler, A. (1998) *Technology and Global Change*. Cambridge: Cambridge University Press.

Hampton, K.N. and Wellman, B. (2001) Long distance community in the network society: Contact and support beyond Netville, *American Behavioral Scientist* 45(3): 477–96.

Hampton, K.N. and Wellman, B. (2003) Neighbouring in Netville, *City and Community*, 2(4): 277–311.

Hanly, M., Dargay, J. and Goodwin, P. (2002) *Review of Income and Price Elasticities in the Demand for Road Traffic, Report to the Dft*, ESRC TSU Publications 2002/13. London: UCL.

Haythornwaite, C. and Wellman, B. (1998) Work, friendship, and media use for information exchange in a networked organization, *Journal of the American Society for Information Science* 49(12): 1101–14.

Hirschman, A.O. (1970) *Exit, Voice and Loyalty: Responses to Decline in Firms, Organizations and States*. Cambridge, MA: Harvard University Press.

Huberman, B.A. and Adamic, L.A. (2005) Information dynamics in the networked world. Working Paper, HP Laboratories, Palo Alto.

Jacobs, J. (1961) *The Death and Life of Great American Cities*. New York: Vintage.

Larsen, J., Urry, J. and Axhausen, K.W. (2006) *Mobilities, Networks and Geographies*, Aldershot: Ashgate.

Levinson, D. and Karamalaputi, R. (2003) Induced supply: A model of highway network expansion at the microscopic level, *Journal of Transport Economics and Policy* 37(3): 297–318.

Licoppe, C. and Z. Smoreda (2005) Are social networks technologically embedded? How networks are changing today with changes in communication technology, *Social Networks* 27(4) 317–35.

Lin, N. (2001) Building a network theory of social capital, in N. Lin, K. Cook and R.S. Burt (eds) *Social Capital: Theory and Research*. New York: Aldine de Gruyter, pp. 3–29.

Low, S.M. (2003) *Behind the Gates: The New American Dream*. London: Routledge.

McCarty, C., Bernard, H.R., Killworth, P.D., Shelley, G.A. and Johnsen, E.C. (1997) Eliciting representative samples of personal networks, *Social Networks* 19(3): 303–23.

McPherson, J.M, Smith-Lovin, L. and Cook, J. (2001) Birds of a feather: Homophily in social networks, *Annual Review of Sociology* 27: 415–44.

Milgram, S. (1967) The small world problem, *Psychology Today* 2(1): 60–67.

Moore, E.G. and Brown, L.A. (1970) Urban acquaintance fields: An evaluation of a spatial model, *Environment and Planning* 2(4): 443–54.

Newman, M.E.J. (2001) The structure of scientific collaboration networks, *Proceedings of the National Academy of Sciences of the USA* 98(2): 404–9.

Newman, M.E.J., Watts, D.J. and Strogatz, S.H. (2002) Random graph models of social networks, *Proceedings of the National Academy of Sciences of the USA* 99 (1): 2566–72, February, 2004.

Ohnmacht, T. (2004) Soziale Netze und persönliche Mobilität: Grundlagen für eine empirische Erhebung, *Arbeitsbericht Verkehrs- und Raumplanung* 250, IVT, ETH Zürich, Zürich.

Pool, I. and Kochen, M. (1978) Contacts and influence, *Social Networks* 1(1): 5–51.

Putnam, R.D. (1999) *Bowling Alone: The Collapse and Revival of American Community*. New York: Simon & Schuster.

Raff, D.M.G. and Trajtenberg, M. (1995) Quality-adjusted prices for the American automobile industry: 1906–1940, *Working Paper* 5032. Cambridge: National Bureau of Economic Research.

Rhode, P.W. and Strumpf, K.S. (2003) Assessing the importance of Tiebout sorting: Local heterogeneity from 1850 to 1990, *American Economic Review* 93(5): 1648–77.

Rumley, P.A. (1984) Amenagement du territoire et utlisation du sol. Dissertation, ORL, ETH Zürich, Zürich.

Schäfer, A. and Victor, D.G. (2000) The future mobility of the world population, *Transportation Research* 34A(3): 171–205.

Scherer, M. (2004) Erreichbarkeitsveränderungen in der Schweiz: Eine kartographische Darstellung, term paper, IVT, ETH Zürich, Zürich.

Schönfelder, S. and Axhausen, K.W. (2003a) On the variability of human activity spaces, in M. Koll-Schretzenmayr, M. Keiner and G. Nussbaumer (eds) *The Real and Virtual Worlds of Spatial Planning*. Heidelberg: Springer, pp. 237–62.

Schönfelder, S. and Axhausen, K.W. (2003b) Activity spaces: Measures of social exclusion?, *Transport Policy* 10(4): 273–86.

Seligman, A.B. (2000) *The Problem of Trust*. Princeton, NJ: Princeton University Press.

Sen, A. (1983) *Poverty and Famines: An Essay on Entitlement and Deprivation*. Oxford: Oxford University Press.

Sobel, J. (2002) Can we trust social capital?, *Journal of Economic Literature* 40(2): 139–54.
Spufford, P. (2002) *Power and Profit: The Merchant in Medieval Europe*. London: Thames and Hudson Ltd.
Stutzer, A. and Frey, B.S. (2004) Stress that doesn't pay: The commuting paradox, *IZA Discussion Papers* 1278. Bonn: Institute for the Study of Labor.
Taylor, P.J. (2004) *World City Network. A Global Urban Analysis*. London: Routledge.
Tönnies, F. (1935) *Gemeinschaft und Gesellschaft: Grundbegriffen der reinen Soziologie*, 8th edn. Darmstadt: Wissenschaftliche Buchgesellschaft.
Travers, J. and Milgram, S. (1969) An experimental study of the small world problem, *Sociometry* 32(4): 425–43.
Tschopp, M., Fröhlich, P. and Axhausen, K.W. (2004) Accessibility and spatial development in Switzerland during the last 50 years: A multilevel regression approach, *Arbeitsberichte Verkehr- und Raumplanung* 260, Institut für Verkehrsplanung und Transportsysteme (IVT), ETH Zürich, Zürich.
Tschopp, M., Sieber, R., Keller, P. and Axhausen, K.W. (2003) Demographie und Raum in der Schweiz, *DISP* 153: 25–32.
Urry, J. (2000) *Sociology beyond Societies: Mobilities for Twenty-first Century*. London: Routledge.
Urry, J. (2003) Social networks, travel and talk, *British Journal of Sociology* 54(2): 155–75.
Van Tilburg, T.G. (1995) Delineation of the social network and differences in network size, in C.P.M. Knipscheer, J. de Jong Gierveld, T.G. van Tilburg and P.A. Dykstra (eds) *Living Arrangements and Social Networks of Older Adults*. Amsterdam: VU University Press, pp. 83–96.
Verhoeven, M., Arentze, T., Timmermans, H.J.P. and van der Waerden, P. (2004) Modeling the impact of key events on long-term transport mode choice decisions: A decision network approach using event history data. Paper presented at the 84th Annual Transportation Research Board meeting, Washington, January 2005.
Vries, J de and Van der Woude, A. (1997) *The First Modern Economy: Success, Failure, and Perseverance of the Dutch Economy, 1500–1815*. Cambridge: Cambridge University Press.
Wasserman, S. and Faust, K. (1994) *Social Network Analysis: Methods and Applications*. New York: Cambridge University Press.
Watts, D.J. (1999) *Small Worlds: The Dynamics of Networks between Order and Randomness*. Princeton, NJ: Princeton University Press.
Watts, D.J. and Strogatz, S.H. (1998) Collective dynamics of 'small-world' networks, *Nature* 393: 440–42.
Wellman, B. (1996) Are personal communities local? A dumptarian reconsideration, *Social Networks* 18(4): 347–54.
Wellman, B. (1999a) *Networks in the Global Village*. Boulder, CO: Westview Press.

Wellman, B. (1999b) The network community: An introduction, in B. Wellman (ed.), *Networks in the Global Village*. Boulder, CO: Westview Press, pp. 1–48.

Wellman, B. (2000) Changing connectivity: A future history of Y2.03K, *Sociological Research Online* 4(4): http://www.socresonline.org.uk/4/4/wellman.html.

Wellman, B. (2001) The persistence and transformation of community: From neighbourhood groups to social networks. Report to the Law Commission of Canada, Toronto.

Widmer, J.P. and Meister, K. (2005) Ausgewählte Schweizer Zeitreihen zur Verkehrs-entwicklung, *Materialien zur Vorlesung Verkehrsplanung* 1, IVT, ETH Zürich, Zürich.

Zahavi, Y. and Talvitie, A. (1980) Regularities in travel time and money expenditures, *Transportation Research Record* 750: 13–19.

Zheng, T., Salganikz, M.L. and Gelman, A. (2005) How many people do you know in prison?: Using overdispersion in count data to estimate social structure in networks. Collective Dynamics Group, Columbia University, New York.

Zhou, W.-X., Sornette, D., Hill, R.A. and Dunbar, R.I.M. (2005) Discrete hierarchical organization of social group sizes, *Proceedings of the Royal Society*, 272B: 439–44.

Chapter 2
Qualitative Interviews on the Formation and Maintenance of Social Networks as Foundation for Quantitative Survey Instruments

Timo Ohnmacht

Introduction

Recognizing the on-going processes of pluralization and differentiation within modern societies, a more detailed investigation of the social realities of travellers is necessary. Chapter 2 presents a qualitative approach to exploring the relationship between the formation and maintenance of social networks and different 'mobilities', as used in recent social science discourse (Urry 2008a).

The analysis is based on 30 in-depth, qualitative interviews conducted in Zurich and Berlin focused on the geographical spread of the interviewees' social networks and the mobilities-driven strategies they use to maintain these networks. These strategies were classified as physical travel (various modes of transport), communicative travel (chats, mobile phone) and virtual travel (travel-relevant information online). The chapter concludes by formulating hypothesis based on these findings to serve as the foundation for the chapters that follow (especially Chapters 3 and 4 in this book).

Theoretical Background on the Formation/Maintenance of Social Networks and Mobilities

Social relations and their spatial distribution are major factors in an individual's concept of space and consequent patterns of mobility behaviour. Within the last two decades, socio-spatial processes have been investigated using the underlying dimensions of 'mobilities' (Urry 2008a). The following forms of 'mobilities' have recently been of major interest in social sciences research (Urry 2000: 49–76):

- *corporeal travel* of people for work, leisure, family life, pleasure, migration, and escape;

- *physical movement of objects* delivered to producers, consumers, and retailers;
- *imaginative travel* through images of places and peoples via television, and other media;
- *virtual travel*, often in real time on the Internet, so transcending geographical and social distance;
- *communicative travel* through person-to-person messages via letters, telephone, fax, and mobile phones.

Numerous sociological profiles of contemporary societies have shown that these forms of 'mobilities' have undergone tremendous change in modern societies. A considerable amount of recent social theory focuses on the inter-relationships between society, 'mobilities' and space. Since Beck (2008); Urry (2008a); Sheller and Urry (2006); Kaufmann et al. (2004) proclaimed a new 'mobilities paradigm' in the social sciences, the analysis of spatial distance and social processes is no longer just an issue for transport studies and transport geography. According to Urry (2008: 54),

> [m]ost social science has not seen distance as a problem or even as particularly interesting (except for transport studies and transport geography). This 'mobilities paradigm', though, treats distance as hugely significant, as almost the key issue with which social life involving a complex mix of presence and absence has to treat.

This 'mobilities paradigm' (Adey et al. 2013) is obviously directly connected to the 'spatial turn' within social sciences (Barney and Arias 2008). Within social sciences after this 'spatial and mobilities turn' in general, geographical space is no longer seen only as a passive container, but also as a relevant factor for enabling and restraining social practices. These authors have argued that space was neglected within the social sciences, but geographical space was already considered relevant in sociological theory as early as the beginning of the twentieth century.

This can be seen especially in the early works of Simmel (1908), who ascertained that spatial distance determines social proximity. Beyond classical sociology, we must also reflect on how the current debates in social science have redefined the connection between social relations and physical space. Recent approaches regard geographical space as a key concept in sociology. Löw (2001) points out that space is shaped by its social, political, cultural and economic conditions. Space should thus be understood as a socially constituted relational concept. On a micro-sociological level, people are actively integrated, constructing space using various ordering processes, thus defining space as a 'relational ordering of living beings and social goods' (Löw 2001). This definition assumes that every social formation produces or reconstructs its specific 'social spatialization' (Shields 1991: 31), which is largely built up by both the places where one lives and works and the locations of social contacts. One can criticize Löw's approach for seeing space primarily as a social construct and for failing to pay much attention to the Euclidean dimension (e.g., physical distance) of space. Nevertheless, a significant amount of social science research has been on the social impact of physical space, such as the challenges posed in trying to maintain long-distance contacts (Latane et al. 1995; Butts 2003); social proximity in one's immediate surroundings (Blake et al. 1956; Caplow and Foreman 1950; Snow et al. 1981); modes of transportation, communication usage and the arrangements of social network geographies (Larsen et al. 2006); and the arrangement of social network geography over the life cycle (Sommer 1996).

How social networks are formed and maintained is shifting. Relationships are weaker, more transient, spatially dispersed and there is less overlap among the various networks to which one belongs (Wellman 1999). There has been much discussion about recent trends in social life, with changes to, or even the end of, social certainties. On a macro scale, Castells describes everything as being in 'flow' (Castells 2001), where in the information age, organizations and much of society as a whole acts on a global scale through ubiquitous communication networks. As a result, the world is shrinking physically, socially and imaginatively. This makes it both broadly possible and even necessary to explore the world. There are social, political, economic and cultural necessities to be 'on the move' (Urry 2000). On a meso scale, Giddens addresses the 'embeddedness' of social certainties in his micro-macro-link sociology; as a result, network members become more individualized with weaker ties (Giddens 1984). In their individualization thesis, Beck and Giddens explore the effects of 'reflexive modern societies' on humans at the micro level (Beck, Giddens and Lash 1994) that lead to a 'mobile risk society' (Kesselring 2008). Consequently, people tend to be more flexible in organizing their life and the so-called 'normal biography' vanishes into spatial diversity and social differentiation (Beck 1986). Social life thus becomes more fluid, dispersed and at a distance, resulting in modern western societies characterized by flows and reflexivity (Lash and Urry 1994). To sum up, globalization on a macro scale and individualization on a micro scale have direct consequences, especially on the quality of social networks.

The connections between mobility and new networking patterns have to be addressed in greater detail when looking at social networks, raising certain questions and issues to be addressed in this chapter:

- What impact does travel behaviour over one's biography have on the maintenance of one's social networks?
- Why is it suddenly necessary for many to travel long distances to meet up with friends, relatives and partners?
- What kinds of 'network tools' (Axhausen 2005; also see Chapter 1 in this book), such as text messages, mobile phones, letters, chats, e-cards, VOIP and laptops with email, are required to maintain relationships?

- What kinds of 'mobility tools' (ibid. 2005; also see Chapter 1 in this book), such as cars, public transport, bicycles and (low cost) airlines, are necessary to meet up with network members in person?
- What causes social networks to become more spatially dispersed over time, consequently influencing our travel behaviour?

This chapter will use this theoretical basis to present a qualitative approach to obtaining first answers to these questions.

Going Qualitative: Research Agenda

The overall objective of the qualitative field research on the formation and maintenance of social networks and 'mobilities' is to gain insights into three main research topics:

- the structure and spatial dimensions of personal social networks (structure);
- life course transitions and their impact upon mobility behaviour related to social networks and 'mobilities' (formation);
- the relationship between mobility patterns and the social network geography (maintenance).

This chapter will present the procedures of this qualitative field study, including the methodological design, the sampling process, the analytical methods to be used, followed up by a critical reflection of the field work.

Methodological Design

Qualitative social research analyses the actions, experiences and reasoning of people based on their comments on a particular subject area (cf. Muhr and Friese 2001). The research design is based on in-depth qualitative interviews. The analysis of these in-depth interviews is based on the researchers' interpretations in light of the theoretical discussion at hand. A key premise of qualitative research is traceability in terms of data collection and analysis; therefore, the methodological design can be summarized as follows:

- *Research question:* The research topic is formulated as an exploratory question.
- *Survey sample*: People are recruited to participate in the qualitative sample.
- *Data collection instruments*: The interview is organized on the basis of questions and subsequently conducted and digitally recorded.
- *Transcription*: The interviews are then transcribed.

- *Explication*: Atlas.TI (a transcription software, see http://www.atlasti.com) is then used to encode individual pieces of text (terms, phrases, passages, …) from the verbal interview transcripts by survey question.
- *Structuring*: In a first step, data material is prepared based on the explication in the form of interview sequences by topic and rooted in a theoretical framework. This creates a keyword-encoded basic data set, which can be processed based on the research question.
- *Summary*: Abstraction is used to create a manageable corpus of data while accurately reflecting the underlying material. This makes it possible to formulate explanations and make statements about certain relationships.

The following sections will provide more in-depth information about the sampling procedure, data collection instruments and analytical methodology. The chapter will then conclude by presenting the key results with a critical methodological reflection.

Sampling

Thirty persons were questioned with the help of semi-structured qualitative interviews during the survey period from October 2004 to February 2005 for their mobility biography, mobility tools and social networks. Overall, information was collected about each interviewee's social network geography as well as their various homes, workplaces and holiday locations over the course of their lives and linked to the qualitative data from the interviews to allow for better interpretation. One-half of the interviews were conducted in each Berlin and Zurich, respectively. Sampling quotas were set to provide the broadest possible dispersion in terms of demographic characteristics such as gender, age, occupation and education. This heterogeneous sampling method was selected to reflect different opinions and attitudes in the qualitative analyses and to allow differentiated insights into the subject matter (strategy of maximum variation, cf. Flick 1995; Lamnek 1995).

Guideline Interview and Data Collection Instruments

The survey instrument consisted of a guideline interview and complementary quantitative survey instruments. The qualitative guideline interviews examined the interviewees' experiences in terms of the spatial dispersion of their social networks, especially the efforts they put in to maintaining relationships both near and far. There was an additional focus on the interviewees' preferred modes of transport and socio-demographic status changes over their lifetimes.

The aims of the complementary quantitative survey was to obtain longitudinal geographical data about complex mobility patterns, such as residential mobility, commuting, travel in terms of social networks and leisure travel. The following quantitative items were asked of each interviewee to measure the size of his or her travel-relevant network geography:

- *Home locations*: surveying the interviewees' residences over their lifetimes.
- *Holiday locations*: surveying the places where interviewees have taken holidays over their lifetimes.
- *Social network*: surveying the home locations of members of the interviewees' social network with whom they have and had close relationships and visit in physical co-presence (and therefore linked to travel; additional information about transport mode most commonly used to visit social network members was also recorded).
- *Workplaces*: The spatial dispersion of workplaces over their lifetimes.

Analytical Method

The analytical method used here is based on elements of grounded theory (Glaser and Strauss 1967) and the evaluation technique of qualitative content analysis (Mayring 1995). In addition, ideal types are presented. Grounded theory is characterized by the recursive linkage of the plan, execution and data analysis within the investigation, which explains why it remains popular particularly in exploratory research contexts and thus finds application here. Grounded theory is based on dialogue and the process nature of 'rolling planning' (Glaser and Strauss 1967). This means that hypotheses are not formulated *a priori*; rather they come about as a theory *grounded* in the data, as the researchers swing between deduction and induction. The *grounded theory* is used to select interviewees for a *theoretical sampling*, i.e. the choice of the next interviewee is based on the experience with the previous interview. This was not the case with these interviews as the breadth of respondents was sufficient to use the analytical methods of *grounded theory*. The 30 interviews ensure sufficient saturation for the research question being considered (cf. theoretical saturation: Glaser and Strauss 1967). The analytical steps were as follows:

- Initially, inductive categories (open encoding) were used to organize the data. The topical categories were generated from the research question itself and applied to the data. This method is based on grounded theory's bottom-up encoding of data material. The selection criteria for the encoded passages is based on the research question, comparable to a deductive element established by theoretical considerations about the object and purpose of the analysis.
- If a text is found that approximates the message of another text passage that has already been encoded, the code is used a second time (subsumption). This is continued until enough categories have been identified to allow an encoding of all passages.
- At this stage, the codes already developed can be used to structure the data material (closed encoding). The transcribed interviews thus form the basis of general concepts in the form of bottom-up codes, which eventually

will be included in the model or theory to be developed about the subject (Strübing 2004).

- Any particularities such as apparent relationships that show up during the encoding process are marked as a memo for future reference in the on-going analysis. The codes and memos are transferred by Atlas.TI to structural trees to allow anticipated connections to be visualized, thus facilitating the analysis. This approach has the overall aim of gaining new insights and answering the research question(s).

Methodological Reflection

The focus of the qualitative interviews is on the respondent's biography in terms of mobility and social network. Biographical research is based on life history data collected retrospectively. Various qualitative interview methods are used to collect reconstructed life histories from the interviewees. The most common method is the narrative interview where individual stimuli are used to elicit autobiographical stories from the respondents (cf. Schütze 1983). Of course, their biographies can never be recorded in full – they are dependent on the reproductive performance of the interviewee (cf. Klein and Fischer-Kerli 2000). Consequently, a distinction is drawn between the actual experience (history) and that which is narrated (story) (Bude 1984). In other words, there is a difference between the life experience and the biography as remembered and told. The mobility biographies presented here is based on the interviewees' reproductive performance and their autobiography as they remember it.

Any empirical analysis is constrained by the limits of the underlying data. The qualitative data used here is biographical information collected retrospectively as part of different projects at different locations, with concomitant consequences on its representativeness. Its validity is based to the group of people selected. Nevertheless, the data is sufficient to examine initial trends that can subsequently be converted into empirically testable hypotheses to investigate statistical correlations within quantitative data sets.

Egocentric network studies describe networks using measures that summarize the relationships of the respondent to network members. Focusing on the interviewees' egocentric networks to determine the connection between their spatial dimensions and their personal mobility biography puts the focus squarely on the dynamics of the individuals themselves. This approach has both its advantages and disadvantages. One advantage is the light shed on events in the interviewees' biography that correlate with the spatial constellation of their networks. This undoubtedly results in detailed insight into those events relevant to the mobility of single individuals *(ego)*. On the other hand, there is a risk of not reflecting on biographical-spatial dynamics of social contacts within the network and thus ignoring the dynamics of significant others (*alters)*. Such simplification must be mentioned in the context of what is possible in view of the data. The data is based on life histories retrospectively collected using the method of *visual history* (cf. Pagenstecher 2003). Since the interviews were mainly conducted in interviewees' homes, reference was often made to supporting documents such as old address books, photo albums and other memorabilia. These would enable details about former acquaintances to be recalled and former preferences for modes of transport and mobility styles to be remembered. It can nonetheless be assumed that there were gaps in the data material. Even short-term travel, transport or mobility diaries show noticeable signs of fatigue (cf. Canzler and Knie 1998: 52), as was also the case with short-term quantitative biographical surveys based on a life course calendar requiring elaborate support (cf. Beige and Axhausen 2003). As this is a fundamental problem of empirical (travel) research (Canzler and Knie 1998: 50), these limitations to data collection must be taken into account.

Relationship between Mobility Biographical Events, Socio-Demographics and Geography of the Social: Exemplary Protagonists and Key Findings

The results of the qualitative interviews are presented with four exemplary protagonists and their mobility biographies, including a comparison between their personal mobility history and their network geographies. In contrast to pure quantitative analyses, this method thus incorporates key problem-oriented self-reflection by the respondents, which can enhance the analytical spectrum for quantitative correlations between social geography and mobility history. The results are thus illustrative and exploratory, while ensuring that the types presented are exemplary ideal cases on the continuum between highly mobile to immobile biographies. In a second section, the key findings from the 30 interviews are summarized into empirically testable hypotheses.

Four Exemplary Protagonists to Enhance the Analytical Spectrum for quantitative Studies

Hyper-mobile Type

Sociological mobility research labels individuals as hyper-mobile if their social and geographical histories are characterized by high mobility (cf. Bonß, Kesselring and Weiß 2004). Claudia has a lifestyle marked with spatial polygamy and her social contacts show a widely spread of spatial distribution.

Claudia's global activity space where she visits and communicates with friends is closely connected to her level of professional mobility as well as her educational biography. She also reports a considerable body of regular long-distance travel. Small portable devices and IT applications such as the mobile phone and laptop allow Claudia to maintain social integration even under conditions, which prevent co-presence. These devices are constant companions during her frequent commutes between London and Zurich as a lawyer for a Swiss insurance company. Her international biography began with her first trip to the USA when she visited one of her parents' business partners. This was followed by a one-year exchange experience at an American high school. Her father, who had come from a northern German city to Zurich to complete his engineering studies at a local university and subsequently met his wife there, felt it was important to maintain contact with his relatives in Berlin. Claudia's first social reference groups and *Lebenswelten* were thus cross-border from the start, based in Switzerland and Germany; visiting relatives remains important to this day. Claudia confirms her high level of geographical mobility from the start of the interview: 'I've always had friends from Germany, Italy and the United States, always replaced [by others] over time'. Her high geographical mobility, which she describes as being the lifestyle 'of a jet-setter' where she might 'just go to Paris for a weekend or to Berlin or Barcelona to visit friends' suggests that her social mobility is also characterized by high geographical mobility: 'It's been my experience that my contacts tend to go dormant and I focus on making new contacts and see them as friends for that particular stage of life and don't spend time searching for long-term friendships that last for years'. After returning from the London offices of the insurance company, it became clear to Claudia in just a few months that she does not see her professional future as being only in Zurich. She puts out feelers for international job openings; and in these 'days of Easy Jet', she has no problem maintaining her social contacts, at least initially after moving away. 'I mostly send emails to people that live far away, but I do exactly the same thing with people who are working right next to me in the office. Nowadays, I actually call people only when I'm late for an appointment or can't find them'.

Figure 2.1 Claudia's activity space

Mobility as Experience-Type

Freedom is what characterizes Tobi's mobility biography. From early on, his openness to mobility was influenced by his father's professional mobility as he travelled the world on behalf of a major German company with his two kids and a wife who helped maintain family life wherever they were living.

Tobi looks at the family home in southern Germany as being an 'anchor point', but family life can happen just as readily in Greece or in Asia, for example. He celebrates family events wherever his father is currently located professionally. Tobi spent three years at a German school in Greece until he was 18 and then decided to settle down in his parents' home in Germany while they were posted to

Figure 2.2 Tobi's activity space

Asia. He began his required civil service and flies several times to Asia for family gatherings with company-sponsored airfares. The familial geography extends to Canada, where his uncle invites Tobi and his family to family gatherings. While his parents were in Asia, Tobi was involved in the local youth centre and organized rock concerts. This political commitment is also what has kept Tobi from obtaining his driver's license: 'My city has the best and most extensive public transport network in Germany. Why does everyone have to drive a Mercedes in Southern Germany?' 'It's typically been the case in my family that we start travelling after finishing school; that's how I landed up in Berlin', referring to his parents' exchange experiences in England and Canada. Changing his residence to Berlin came about as a result of Tobi's lifestyle. He is formally enrolled in a Chinese Studies programme, reflecting an interest in Asia obtained from visiting his parents, but he is actually pursuing his main interest in Berlin: booking bands in Berlin's trendy venues. He is quickly building a network of other young people (professionally) interested in music in Friedrichshain-Kreuzberg, a neighbourhood he singles out for its being a place for 'leisure and fun'. As a roadie on the European tours of several bands, he has also met people with whom he keeps in contact mainly by email to set dates to host the bands handled by his newly formed booking agency. Tobi is also often on the road to Scandinavia, home to many of the bands signed by his booking agency. After three semesters of Chinese Studies, he begins an internship at a small media design firm in Prenzlauer Berg. He then enrols in Cultural Studies at the Viadrina University in Frankfurt an der Oder. With the Brandenburg commuter pass in hand, he commutes the 40 minutes from Berlin to Frankfurt to 'whatever' seminars happen to strike his fancy. He takes courses that offer DAAD (German Academic Exchange Service) funding for research on football in Brazilian favelas. As a result of such stays, his personal mobility biography includes overseas stays that last for several months at a time. He also uses online carpooling sites to organize affordable trips to visit friends in such places as Zurich, Hamburg and Vienna. He tries to combine his studies, friendships and his music-booking agency. He follows a Buddhist belief system and tries to free himself from material things and focus his life on the essentials: 'What I need is an iPod, because I can fit all my music on something the size of a chocolate bar; I've sold all my CDs and records on eBay. The same with my books; you can only read one at a time anyway. What I also need is my Samsonite suitcase with my favourite clothes, my Apple PowerBook and a flat rate mobile plan, as well as a sub-lease for a small room in a Berlin apartment so that I'm not held down by a long-term rental agreement. My life motto is: be ready to move in 15 minutes and be ready to visit a friend in another city within 5'. He sees the internet as a geographical space 'www in front of it'. Telepresence allows relationship functions that cannot often be done in person due to spatial mobility constraints to be substituted by functional equivalents: Skype, email and online communities.

Mobile-immobile Type

Kristof would be labelled in sociological mobility research as mobile-immobile (see Bonß, Kesselring and Weiß 2004). He travels on business for his employer, a Swiss TV channel, and has maintained his key contacts in his home cantons of Zurich and Aargau since his earliest days: a weekly phone call with his mother at a fixed time is an absolute must for Kristof!

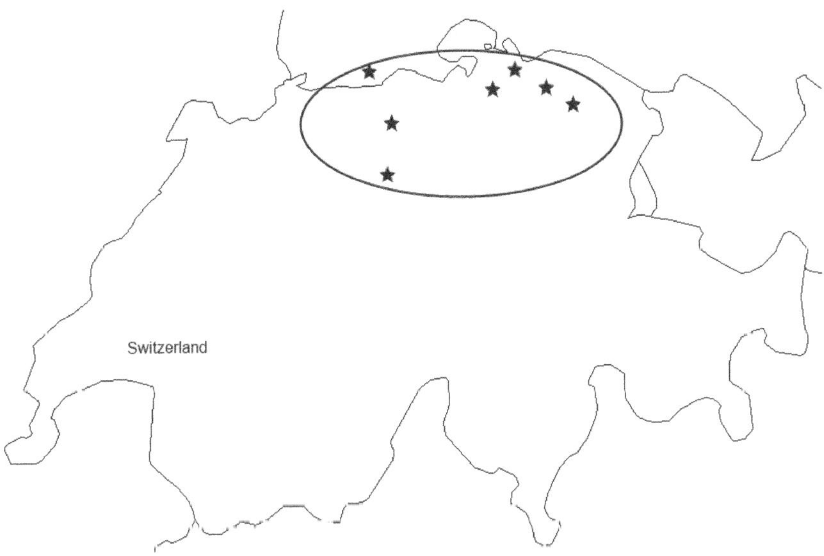

Figure 2.3 Kristof's activity space

Like products and capital, the work for Swiss television has also become more mobile. Various locations have become more important as has the demand for mobile workers. Thus Kristof finds himself commuting between Russia, the Caribbean, Canada and Switzerland to do television reports, mostly holiday programmes. Kristof's most important means of transport is his car, which he used, for example, to meet one of his two sons in southern Germany: 'For me, a lot depends on my work. If I'm doing some outside product, I always arrange a visit. If I head out to Lake Constance, I meet with him. That's how these meetings usually get arranged'. 'For me, it always comes down to the car. My father was a chauffeur. Even when I was a kid, we didn't travel any other way. My parents never took the train; they only travelled by car'. 'When I was still living with my parents in Ticino, they still honked when a German-speaking Swiss drove past'. While Kristof no longer does that, but, like his parents, he keeps an extremely small social activity space in his private life: 'What others do in their free time is what I do for a living: I travel, I stay in hotels, I go to concerts and events, I eat out, all professionally. That's why my social network is based on my job. My private social network actually only includes my direct neighbours'. Kristof's socialization takes place on the job. He even reports how retired production managers and editors will still come to the Swiss television canteen to hang out with their 'work friends'. Kristof's example shows that the workplace operates on a different timeframe than one's home. There are types of work where spatial and temporal conditions are extensively self-determined: Kristof has a global professional network, but in his private life moves physically barely outside of his immediate geographical environment.

Local Type

Werner has moved three times in his life. Any changes to the centre of his life have been the result of work-related changes. Having been granted a residence permit for East Berlin, as was required at the time for GDR citizens, he moved from a provincial Brandenburg village to East Berlin, living first in Hellersdorf and then in Marzahn. His 'job as construction manager' led to his working on the construction of the GDR Palace of the Republic in central Berlin. 'My mother never got further than Bochum to visit her relatives', he says, summing up his background. 'Maybe she went once or twice to Leipzig or Berlin to go shopping'.

His family had always lived in Fürstenberg, since 1450. He spends his holidays in the immediate area, the Mecklenburg Lake District. Later, he sometimes goes to the Baltic or to the Erzgebirge within East Germany with his family: 'We couldn't afford anything more'. After a herniated disc ends his career, Werner opens a shop in Friedrichshain in 1995, where he sells exotic Asian swords, among other things. At the age of 50, he also fulfilled a long-held dream: to take his first long-distance trip to Japan: 'For me, the craftsmanship in those countries is most important, since I come from a crafting profession'. In addition to visits to his home village, he also arranges by telephone to visit with friends in East Berlin, travelling on Berlin's public transport system.

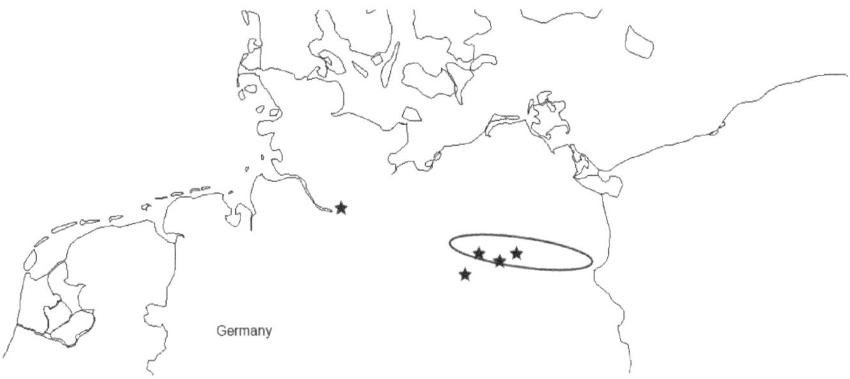

Figure 2.4 Werner's activity space

Summary of Key Findings

The 30 transcribed interviews were examined for the following issues, based on the three research questions on the structure, formation and maintenance of social networks and their relationship to mobilities:

- the importance of proximity in long-distance relationships (maintenance);
- the influence of parents on individual mobility; such as stories of second-hand experiences passed down in nuclear families and the influence of parents' travel behaviour on the following generation (formation);
- how communication forms and modes of travel are utilized to maintain the social networks (maintenance).

The main interest is in locating events, i.e. situations, which affect mobility behaviour. Interviewees were consequently asked to put these events into the context of their personal mobility behaviour. The following exemplary life transitions affecting personal mobility patterns were identified in the 30 interviews:

- *relocation* – moving from one place to another involves a new analysis of the local infrastructure, changing the routines of mobility patterns.
- *flexibility* – for example, people caring for persons with disabilities tend to acquire cars to obtain more flexibility.
- *mobility tools* – access to mobility tools in general, for example acquiring a driver license is a life course transition per se.
- *household structure* – the birth of children is another life course transition for the parents. New mobility patterns evolve, for example, fewer holidays. Young families tend to buy a car to organize their life more easily.
- *critical incidents* – negative experiences with mobility tools, such as accidents, also influence mobility patterns.

- *positive incidents* – positive experiences with mobility tools, such as the first flight with a low-cost airline influence mobility patterns.
- *ageing* – age-related immobility with the use of support to maintain personal mobility.

The following items offer some insight into the influences of social networks upon patterns in individual mobility biographies:

- Long-distance social contacts are associated with the choice of holiday destination; many interviewees mentioned holidays as a chance to combine holiday-making and visiting friends.
- Most of the interviewees' network geography increases over their lifetime, with the exception of those who have a dispersed familial geography, as often found in diasporic cultures.
- Special occasions such as weddings, birthdays, funerals, stag nights are reasons for long-distance meet-ups.
- Low-price communication supports long-distance social contacts, for example, Deutsche Telekom's 'Free Sunday' tariff and VOIP technologies (such as Skype).
- Generally low transport costs also support the maintenance of long-distance social contacts. Low-cost airlines have proven helpful in maintaining long-distance social contacts, according to interviewees. Strong ties are not necessarily related to living nearby.
- As the number of events in the mobility biography increases, the social network becomes more spatially dispersed. For example, people with high job mobility and changes in residences have more dispersed personal networks, which may result in a complex work-life balance with elaborate personal mobility management.

There are also some interesting inter-generational patterns in mobility biographies:

- Children accept and reject their parents' mobility patterns in different ways. Some children develop different mobility patterns at various life stages.
- Most of those interviewed described their favourite landscape as that associated with the parental home.
- Adult children's holiday frequency and destinations are related to their parents' holiday frequency
- And parents and children have similar mobility tool preferences.

Final Considerations: Hypotheses for a Quantitative Field Study

The heterogeneity of the four exemplary cases is positive, because it enables differentiated insights into different demographic characteristics such as gender, age, education and occupation. This is particularly advantageous within qualitative research as it takes into account different opinions and attitudes on mobility behaviour (strategy of maximum variation; cf. Flick 1995). The transformation processes, both technological and social, in the development of modern society addressed in the theoretical section (such as the development of accessibility through innovations in transportation technology and infrastructure, biographical projection through increased mobility and individualization and changes in communication technology) are brought from the level of abstraction back to the individual. The theoretical framework of this chapter allows the following key points to be made:

- social networks have become larger and more extensive;
- people participate in different social spheres, increasing the complexity of social networks;
- the social networks are undergoing a major increase in variability. Biographies are increasingly characterized by caesura and need for self-development.

The overall goal, in conjunction with the chapters to follow (especially see Chapters 3 and 4 in this book), is to make a contribution to empirical quantitative research on the spatial distribution of social relations and to discuss the empirical evidence on the spatial organization of social relations given current research in transport and mobility. The focus is on the following questions:

- What is constitutive of the social profile of those who care about a large number of contacts that go beyond the local?
- What kinds of transportation, mobility tools are used to maintain these supra-local contacts?
- In what context is the desire to spend more time with friends, global life satisfaction and mobility understood as enrichment, being connected with one's community, community size with close supra-local contacts?
- What role does age play in the spatial geography of contacts of people in single-person households, especially since research shows that older people are more likely to need help from those in the immediate vicinity (Ohnmacht 2009)?

The following hypotheses can be derived with reference to the chapters that follow. Gender-based differences need to be assessed to determine whether women maintain more compact network geographies than men due to the continued strong division of gender roles with stronger ties to local activity spaces. For example, Löw (2001) points out that the majority of girls are socialized to activity within a reduced spatial sphere, whereas boys learn to expand their sphere of activity (246 ff.). Similarly, it is hypothesized that high income and a correlated level of post-secondary education as well as an age of up to 29 years can be connected with spatially extended networks, since these are traits that describe people who are more mobile and who have more mobile networks (for Germany, cf. Kesselring 2006). According to Simmel (1908), spatial fixations can be marked by where social contacts reside. There is empirical evidence from migration research (cf. Grieco 1995). The places associated with social relationships are considered potential destinations for emigration (Beck 1997: 58). An important aspect is the interaction of migrants between the place of origin and the place of emigration, as seen in various research approaches (Espinosa and Massey 1997; Fall 1998; Beck-Gernsheim 2007). This qualitative research shows that social relationships create a strong attachment to places of origin even after emigration. People with a migration background do not simply leave the places of their origin behind, unless occasional visits are rendered impossible by war or for political reasons. If this is not the case, weddings, Christmas and other occasions are often reason to return to the place of origin. Another hypothesis to be tested for Switzerland is whether people active in organizations or in politics have their close contacts primarily in close vicinity. Because engagement in organizations or politics can serve as an indicator of community cohesion and a socially-active lifestyle within one's close vicinity. Simmel (1908) points out that social communities can manifest material fixations in space. For example, for religious communities, this can be expressed by the church that stands at the centre of its members' activities (Frisby 1984: 128). Also, membership in groups with shared interests, which is accompanied by social obligations and checks, can result in a spatial fixation. This then, according to the thesis, can result in a strong local social integration, which leads to the fourth hypothesis: the quantitative empirical studies should specifically address the direct connection between the events described in mobility biographies and the individuals' number of active contacts and the spatial distribution of their social networks. The number of active contacts and their spatial structure can have a decisive influence on people's mobility patterns. The homes of relatives, partners and friends can be viewed as 'touchstones' in socially-conditioned mobility patterns. These touchstones set out a socio-geographic space within which a person maintains his or her contacts. To examine the influences of spatial flexibility and stability of an actor on the spatial distribution of the social network, biographical events should be empirically examined for the spatial distribution of social relations. There is an empirical objective to find out whether close social relationships over long distances are correlated with a need for more frequent personal visits to determine the social stress caused by time-consuming trips to maintain social relations. It must also be examined whether spatial distance to close contacts is associated with life satisfaction in general. It is also of interest whether life satisfaction and quality of life can be associated with whether close contacts live nearby. Also concerning these two hypotheses, one must examine whether people consider mobility less as an enrichment when close contacts are spatially distant and consequently seen less frequently. It can therefore be assumed that a person closely connected to his or her home is also well integrated with close local relationships. Table 2.1 summarises the hypotheses which will mostly be addressed in the following chapter.

Table 2.1 Hypothesis

1	Women have spatially more compact network geographies than men.
2	People with high income and post-secondary education have spatially more expansive networks.
3	Young adults have spatially more expansive networks.
4	People with a migration background are characterized by spatially more expansive social networks.
5	People in locally-based interest groups, such as clubs or political associations, have spatially more compact social networks.
6	The more eventful a person's mobility biography, the spatially more dispersed is her or his social life.
7	The more eventful a person's residential biography, the spatially more dispersed is her or his social network.
8	The more eventful a person's educational biography, the spatially more dispersed is her or his social network.
9	The more eventful a person's professional biography, the spatially more dispersed is her or his social network.
10	People want to spend more time with friends, when emotionally close contacts are spatially distant.
11	Life satisfaction is higher when close contacts live in closer proximity.
12	Persons with distant close contacts consider mobility less frequently as enrichment.
13	A person connected to his or her home town has many close local contacts at the current residence.
14	The size and spatial extent of the social network is different according to socio-demographic characteristcs and the mobility biography.
15	How people stay in contact with their social network (face-to-face meetings, Email, SMS etc.) is associated with the size and spatial extent of the social network, socio-demographic characteristcs and the mobility biography.

References

Adey, P., Bissel, D., Hannam, K., Merriman, P. and Sheller, M. (eds) (2013), *The Routledge Handbook of Mobilities*, London: Routledge.

Axhausen, K.W. (2005), Social aspects of sustainable transport: Transatlantic perspectives. In: K.P. Donaghy, S. Poppelreuter and G. Rudinger (eds) *Social Dimensions of Sustainable Transport: Transatlantic Perspectives*, Aldershot: Ashgate, pp. 90–109.

Barney, W. and Arias, S. (eds) (2008), *The Spatial Turn: Interdisciplinary Perspectives*, London: Routledge.

Beck, U. (1986), *Risikogesellschaft*, Frankfurt/Main: Suhrkamp.

Beck, U. (1997), *Was ist Globalisierung? Irrtümer des Globalismus – Antworten auf Globalisierung*, Frankfurt/Main: Suhrkamp.

Beck, U. (2008), Mobility and the cosmopolitan perspective. In: W. Canzler, V. Kaufmann and S. Kesselring (eds) *Tracing Mobilities: Towards a Cosmopolitan Perspective*, Aldershot: Ashgate, pp. 25–35.

Beck, U., Giddens, A. and Lash, S. (1996), *Reflexive Modernisierung: Eine Kontroverse*, Frankfurt/Main: Suhrkamp.

Beck-Gernsheim, E. (2007), Transnational lives, transnational marriages: A review of the evidence from migrant communities in Europe, *Global Networks* 7(6): 271–88.

Beige, S. and Axhausen, K.W. (2008), Long-term and mid-term mobility decisions during the life course: Experiences with a retrospective survey, *IATSS Research* 32(2): 16–33.

Blake, R., Rhead, C., Wedge, B. and Monton, J. (1956), Housing architecture and social interaction, *Sociometry* 19(2): 133–9.

Bonß, W., Kesselring, S. and Weiß, A. (2004), Society on the move: Mobilitätspioniere in der Zweiten Moderne. In: U. Beck and C. Lau (eds) *Entgrenzung und Entscheidung*, Frankfurt/Main: Suhrkamp, pp. 258–79.

Bude, H. (1984), *Rekonstruktion von Lebenskonstruktionen – eine Antwort auf die Frage, was die Biografieforschung bringt*, Stuttgart: UTB.

Butts, C. T. (2003), Predictability of large-scale spatially embedded networks. In: R. Breiger, K. Carley and P. Pattison (eds) *Dynamic Social Network Modelling and Analysis: Workshop Summary and Papers*, Washington D.C.: National Academies Press, pp. 313–23.

Caplow, T. and Foreman, R. (1950), Neighborhood interaction in a homegenous community, *American Sociological Review* 15(3): 357–66.

Castells, M. (2002), The space of flows. In: I. Susser (ed.) *The Castells Reader on Cities and Social Theory*, Oxford: Blackwells, pp. 314–66.

Espinosa, K. and Massey, D.S. (1997), Undocumented migration and the quantity of social capital, *Soziale Welt* 12 (Special Issue): 141–62.

Fall, A.S. (1998), Migrants' long-distance relationships and social networks in Dakar, *Environment and Urbanization* 10(1): 135–145.

Flick, U. (1995), *Qualitative Forschung*, Reinbeck: Rowohlt.

Flick, U. (2000), Triangulation in der qualitativen Forschung. In: U. Flick, E. v. Kardorff and I. Steinike (eds) *Qualitative Forschung. Ein Handbuch*, Hamburg: Rowohlt Verlag, pp. 42–65.
Frisby, D. (1984), *Georg Simmel*, London: Tavistock.
Giddens, A. (1984), *The Constitution of Society: Outline of the Theory of Structuration*, Cambridge: Polity Press.
Glaser, B. and Strauss, A. (1967), *The Discovery of Grounded Theory. Strategies for Qualitative Research*, New York: Aldine.
Grieco, M. (1995), Transported lives: Urban social networks and labour circulation. In: S. Vertovec and A. Rogers (eds) *Explorations in Anthropology Series*, Oxford: Berg, pp. 36–50.
Kaufmann, V., Bergman, M.M. and Joye, D. (2004), Motility: Mobility as capital, *International Journal of Urban and Regional Research* 28(4): 745–765.
Kesselring, S. (2006), Topographien mobiler Möglichkeitsräume: Zur sozio-materiellen Netzwerkanalyse von Mobilitätspionieren. In: B. Hollstein and F. Strauss (eds) *Qualitative Netzwerkanalyse. Konzepte, Methoden, Anwendungen*, Opladen: VSA Verlag.
Kesselring, S. (2008), The mobile risk society. In: W. Canzler, V. Kaufman and S. Kesselring (eds) *Tracing Mobilities: Towards a Cosmopolitan Perspective*, Aldershot: Ashgate, pp. 77–102.
Klein, T. and Fischer-Kerli, D. (2000), Die Zuverlässigkeit retrospektiv erhobener Lebensverlaufsdaten. Analysen zur Partnerschaftsbiographie des Familiensurveys, *Zeitschrift für Soziologie* 29(4): 294–331.
Knie, A. and Canzler, W. (1998), *Möglichkeitsräume: Grundrisse einer modernen Mobilitäts und Verkehrspolitik*, Köln: Böhlau Verlag.
Lamnck, S. (1995), *Qualitative Sozialforschung*, Weinheim: Beltz PVU.
Larsen, J., Urry, J. and Axhausen, K.W. (2006), *Mobilities, Networks, Geographies*, Aldershot: Ashgate.
Lash, S. and Urry, J. (1994), *Economies of Signs and Space*, London: Routledge.
Latane, B., Liu, J.H., Nowak, A., Bonevento, M. and Zheng, L. (1995), Distance matters: Physical space and social impact, *Personality and Social Psychology Bulletin* 21(8): 795–804.
Löw, M. (2001), *Raumsoziologie*, Frankfurt/Main: Suhrkamp.
Mayring, P. (1995), *Qualitative Inhaltsanalyse: Grundlage und Techniken*, Weinheim: Deutscher Studienverlag.
Muhr, T. and Friese, S. (2001), Computerunterstützte Qualitative Datenanalyse. In: T. Hug (ed.) *Wie kommt Wissenschaft zu Wissen?* Band 2: *Einführung in die Forschungsmethodik und Forschungspraxis*, Balltmannsweiler: Schneider Verlag, Hohengehren, pp. 380–400.
Ohnmacht, T. (2009), Social-activity: Do the 'strong-tie relationships' of a person exist in the same community? The case of Switzerland, *Environment and Planning* 41(12): 3003–22.

Pagenstecher, C. (2003), *Der bundesdeutsche Tourismus: Ansätze zu einer Visual History: Urlaubsprospekte, Reiseführer, Fotoalben, 1950–1990*, Hamburg: Verlag Dr. Kovac.

Schütze, F. (1983), Biografieforschung und narratives Interview, *Neue Praxis* 3(6): 283–93.

Sheller, M. and Urry, J. (2006), The new mobilities paradigm, *Environment and Planning A* 38(2): 207–26.

Shields, R. (1991), *Places on the Margin: Alternative Geographies of Modernity*, London: Routledge.

Simmel, G. (1908), Der Raum und die räumliche Ordnung der Gesellschaft. In: G. Simmel (ed.) *Soziologie: Untersuchungen* über *die Formen der Vergesellschaftung*, Berlin: Duncker und Humblot Verlag, pp. 460–526.

Snow, D.A., Leahy, P.J. and Schwab, W.A. (1981), Social interaction in a heterogeneous apartment: An investigation of the effects of environment upon behaviour, *Sociological Focus* 14(4): 309–19.

Sommer, R. (1996), *Personal Space*, Englewood Cliffs: Prentice Hall.

Strübing, J. (2004), *Grounded Theory*, Wiesbaden: VS Verlag.

Urry, J. (2000), *Sociology beyond Societies: Mobilities for the Twenty-First Century*, New York: Routledge.

Urry, J. (2008), *Mobilities*, Oxford: Blackwell.

Wellman, B. (1996), Are personal communities local? A dumptarian reconsideration, *Social Networks* 18(4): 347–54.

Chapter 3
Egocentric Networks in Zurich: Quantitative Survey Development, Data Collection and Analysis

Andreas Frei and Timo Ohnmacht

Introduction

This chapter addresses the size and structure of egocentric social networks. It is theoretically based on the sociology of space, mobility studies, and debates on social capital. Based on Chapter 2 the research findings and hypotheses are used to address the structure of egocentric social networks in five parts:

1. number and configuration of relationships in social networks,
2. use of information and communications technology (ICT),
3. respondents' mobility biography,
4. mobility tools and
5. sociodemographic attributes.

A representative survey has been designed and conducted to investigate the above listed five parts of social network structures. The data from this survey will be used throughout this chapter as the basis for a quantitative analysis of the relationship between social network, geography, travel and mobility biography.

Sampling Method and Survey Instrument

The scope of the survey was derived from initial qualitative work (Ohnmacht and Axhausen 2005; Larsen, Urry and Axhausen 2006) described in Chapter 2, related quantitative work on mobility biographies (Beige and Axhausen 2006) and a substantial pre-test (Frei 2007). The survey instrument addresses:

- the basic socio-demographics of the respondents,
- the mobility biography of residential and employment changes over the lifetime, including information about income levels, mobility tool ownership, main mode of transport to work and

- the ego-centric network through two name-generator with four prompts and a name interpreter, which establishes among other items the exact home location of the contacts (alters) of the respondents and the frequency of their interactions by four modes: face-to-face, phone, email and texting (short-message-service – SMS).

This unique combination reflects the conviction that the social networks depend on the life course of the respondent, as interactions with friends, relatives and others do not depend as much as in previous decades on their physical vicinity. The social network may accumulate over the places the respondents have lived and worked. The survey permits a range of quantitative analyses to test the qualitative hypotheses discussed in Chapter 2, mainly the size and structure of social networks in the context of travel and geography.

Survey Design

The scope of the survey outlined above adds mostly items to the person description through the mobility biography and the social-network questions. Name generators and mobility biographies are a substantial challenge as the respondents' willingness to cooperate is expected to be lower the more intrusive the questions are. The issue of the response burden is another challenging issue raised in the literature and will be discussed in the following sections.

Mobility Biographical Questions

While it is possible to build on the existing literature, the first experiences within the field of transportation planning show the difficulties with social-network and mobility biography survey elements. To capture the mobility biography, Lanzendorf (2003) and Ohnmacht and Axhausen (2005) used a year by question grid in a face-to-face interview, which was filled out by or with support from the interviewer. Beige and Axhausen (2006) employed a grid in a postal survey, following examples in the literature (see Kluge and Kelle 2001). The quality of the responses was good, but the response rate was rather low. Larsen et al. (2006) replaced the grid with independent lists, which seemed to be easier for the respondents, many of whom regarded grids as difficult. Schiffmann (2005) also turned to lists, which he found were filled in well. For this questionnaire a list style was used as well following the previous survey designs. These single lists were collected continuously as retrospective data along major life anchor-points or adjustments over the entire life course. Firstly, it concerned important events of the personal and family history, such as the move from the parental home, marriage, divorce, births, deaths, retirements, etc. These events were used as anchor-points to prompt the memory of the respondent, to facilitate the reconstruction of the life course and to increase the accuracy of the remembered information (Bruckner 1990). Second, the existing and former residences and workplaces were surveyed with their geographical details. In addition, the possession of mobility tools in the form of the availability of cars and possession of various public transport season tickets were asked. As a last item, the respondents were interviewed about their memberships in clubs and the nature of the association.

Name-Generator

Problems with name generators are well known: conveying the type of social contact, which is of interest to the study to the respondents; the trade-off between the respondents' burden and the number of contacts retrieved; and the level of detail of the contact descriptions (for a comprehensive review, see Marsden 1990 and 2005). The type of contact of interest will vary with the study: some focus on relatives, some on physical neighbours, some on fellow students or employees, and some on friends. But the studies have the personal network approach in common, which focuses on specific individuals (egos) and their social contacts (alters). The number of contacts a person has is estimated to range into the thousands (see e.g. Pool and Kochen 1979). To define the network boundaries, usually a set of name generating questions is used. Different name generating approaches can be distinguished (Molin et al. 2008), each of them capturing a certain portion of the entire network. Four main approaches can be distinguished. The interaction approach asks for a record of all alters with whom the respondent (ego) interacts in a certain period of time (including casual and unknown contacts). Another approach is the role relation approach: a record of people with whom the individual has a certain role relationship, such as immediate family, relatives, neighbours or friends. Thirdly, the affective approach asks respondents to record the people with whom they have a close personal relationship, or who are especially important to them. Finally, the exchange approach concentrates on people with whom the individual has social exchange, such as help, social activities or talks about worries. In general, once all alters are elicited, additional questions are used to gather more information on the characteristics on each alter and the ego-alter relationship. These additional questions are called name interpreters. Carrasco et al. (2008); Wellman et al. (2005) for example used an affective approach. Respondents were asked to name the persons who live outside their household, with whom they felt very close or somewhat close. Very close people consist of those persons with whom the respondent discusses important matters or regularly keeps in touch with, or are there for them if they need help. Somewhat close people were described as those persons who are more than just casual acquaintances, but not considered to be very close.

None of the approaches above really focuses on those contacts, which have an impact on the respondents' travel. It is apparent that comprehensive, or at least very long, listings can only be retrieved from written records of some kind, including personal sources such as e-mail archives, telephone bills, address books, telephone numbers stored on mobile phones, Christmas mailing lists, and publications lists; and institutional sources, such as yearbooks, membership lists, internal telephone directories, and minutes of meetings. A comprehensive fusion of all relevant administrative sources and diary data has not yet been attempted due to budget and response burden constraints. Still, the core of the personal networks should be retrievable if Aiello and Dunbar (1993) and Zhou et al. (2005) research findings are correct. They suggest that the core should have five to fifteen members, of which only three to five are the central reference group. The problem, as observed above, is that this central group might not be relevant to some of the travel the analyst is interested in. These journeys are often to contacts, which are strong enough to host the respondent, but are not necessarily to members of the core group. A survey strategy, which tries to amass a large number of contacts is likely to backfire, as the attendant response burden increases the likelihood of systematic survey non-response. Starting from the rich experience of sociology with name generators, a name-generator appropriate to a stimulus was chosen. With a stimulus a certain kind of activity is given, e.g. discussing important matters, for which the interviewee names alters. With this approach the possible lack of clarity in the understanding of specific social contexts such as friendship can be avoided. Name-generators and name-interpreters together are the network-instrument. The best known and used instruments are Burt's (Burt 1984), Fischer's (Fischer 1982) and Wellman's instrument (Wellman 1979). Burt's instrument was used for the first time 1985 in the General Social Survey (GSS). A random sample of 1,534 persons was interviewed with the following name-generator: 'From time to time, most people discuss important personal matters with other people. Looking back the last six month – that would be back to last August – who are the people with whom you discuss an important personal matter?' (Burt 1984). Fischer's instrument was designed for Northern California Community Study (NCCS). 1,050 persons were asked with Fischer's instrument, which uses ten questions appropriated to stimulus. Fischer's name generator (Fischer 1982: 290) asks for the first names of those who:

1. respondents would ask to look after their homes when they go out of town:
2. had helped with tasks around the house in the previous three months:
3. they talk with about how they do their jobs (asked only of employed respondents):
4. they do various social activities with – sharing a meal, visiting. going out socially, etc.;
5. they get together with to talk about hobbies;
6. they date seriously or consider getting married (asked only of unmarried respondents);
7. they talk with about personal matters of concern;
8. they rely on for advice about important decisions; and
9. they would or could ask for a sizable loan.
10. 'Is there anyone who is important to you who doesn't show up on this list'?

Coates and Wellman obtained ego-centric social networks 1968 in East York, Toronto. As Burt they used only one name-generator: 'I'd like to ask you a few questions about the people outside your home that you feel closest to; these could be friends, neighbours or relatives' (Wellman 1979). Only the first six of the named alters were recorded and described in detail. For the research goal here of measuring the size and structure of social network geographies influencing travel behaviour, destination choice or residential choice, it is not so important to survey alters from all different social context spheres but to obtain ego's most important alters. Therefore we used an adapted and appropriate set of name-generators as stimuli. The respondents were handed two lists with two different name-generators. The first name generator is adapted from Burt's and Fischer's instrument where we asked for persons with whom the respondents 'discuss important problems, with whom you stay in regular contact or which you can ask for help'. These questions cover the ten stimuli from Fischer in an abstract way as Burt's generator does and should cover the 'very close' or 'most important' core contacts. The second name generator asked for persons with whom the respondents spend and plan leisure time. This generator targets less strong ties against the background that leisure travel makes up the largest share of long distance travel. The name-interpreter asked then for all of the named contacts how they got to know each other, how long the relationship exists, the frequency of contacts by different modes (face-to-face, telephone, email and SMS – short message service via mobile phone), where they met the last time and the contact's place of residence. The list of modes was left incomplete by design, as we assumed that letter writing has become marginal and that computer-supported chat would follow the patterns of other asynchronous modes (email, SMS). There was also concern about the total response burden. The origin of the acquaintance was categorized as family, subdivided in first degree, relatives or partner, from/of work, education or partner or others. The frequency of contacts should be specified as accurately as possible e.g. every week, 2 times per year. The contact's place of residence should be indicated with as much detail as possible (post code, municipality, street and house number).

The individual questionnaire included general questions about age, gender, nationality, education or employment (nature, scope, location) and for possession of a mobility tools. The survey can be found in Frei (2012).

Survey Protocol Design

Given the difficulties in winning the respondents' cooperation for even simple surveys, the range of questions listed above, which add to the usual, quite numerous questions, raises the problem of response burden. The rather personal nature of the questions about social networks does not simplify the issue. The face-to-face interview is the best method of establishing the trustworthiness of the interviewer and, by extension, the study. It would be possible to include an incentive to the protocol; however, this might become large enough to raise worries about the motivation of the participants. During a pre-test (Axhausen, Frei and Ohnmacht

2006 and Frei 2007) three different survey methodologies (only self-completion; only face-to-face, mixed face-to-face and self-completion) were tested to identify a survey format which minimized missing values due to fatigue effects and reduced the recall problems of the retrospective survey items. The mixed method approach was the best approach with an acceptable cost per response (110 CHF/ usable response[1]).

Field Work

The main study adopted the mixed face-to-face protocol tested in the pre-test. For the survey, 4,200 Zurich city residents with available addresses and telephone numbers were chosen randomly. Following an announcement letter, the subjects were contacted on different days of the week and times of day, and then recruited during the telephone interview, including arranging appointments for the face-to-face interviews. With the reminder notice for the interview, respondents received the written form allowing them to raise questions during the upcoming interview. The written part consists of a person form and a form with mobility biographical questions about relocations, former and current job locations, usage of mobility tools, important life events and memberships in groups that meet periodically. The one to two hours long face-to-face interview covered the social contact questionnaire, but was also used to detect and address respondent difficulties and to establish rapport with the respondent.

Response Rate in the Main Study

The main study aimed to collect 300 complete surveys. From a population of about 100,000 purchasable addresses representing the population of persons over 18 years old living in the city of Zurich, a gross sample of 4,200 persons was drawn. The net sample size contained 2,714 people (64.4%), after deducting the households not reachable by telephone within five attempts and false addresses. The recruitment was based on a pure probability sampling. Finally, of these 332 persons were recruited, of whom 307 (11.3%) were interviewed and completed the questionnaire. The interviewees received no incentive. See Table 3.1 for more details.

1 About US$ 110 at the 2007 rates.

Table 3.1 Response rate for the 2007 Zürich survey on social networks and mobility biography

Phase	Pretest	Main study	Share [%]	% reached by phone
Sample	150	4,200	100	
Incorrect address	0	56	1	
Not reached by phone	36	1,486	36	
Reached by phone	113	2,714	65	100
Recruited	14	318	8	12
Interviewed	13	305	7	11
Written questionnaire returned	13	298	7	11

Source: Axhausen and Frei (2007).

Representativeness of the Data Collection

Table 3.2 shows the socio demographic characteristics of the respondents in comparison with the Zurich population, as observed in the Swiss Microcensus Travel 2005 (ARE and FSO 2007) and the Swiss Census 2000 (FSO 2000). The income information is not directly comparable because the Microcensus measures the household income while this study is person-based. The comparison shows that the Zurich sample is a little bit older, slightly better educated and has a higher share of public transport season tickets. The mean age for the respondents with a university degree is with 43.61 years below the population mean and for those with an obligatory schooling higher with 56.77 years respectively 59.66 years for those with a vocational training. Respondents with no public transport season tickets are over-represented in the group of people with vocational training and a university degree. Based on the sample size, the deviation in the sample from the population is within the expected range. For e.g. the share of male people in the sample, the standard deviation is 0.028. For a confidence interval of 95% the t-value is 1.96 and therefore the confidence interval from the sample is 43.6+/−5.54, which includes the population mean. Overall a re-weighting of the data seems not necessary given the relatively small deviations. It is also likely that the sample from a total of around 300 individuals provides enough variance for the subsequent multivariate analyses. Because the goal of the analysis is primarily to verify causal hypotheses, the extrapolation of the values of individual indicators on Zurich's whole population is of secondary importance.

Table 3.2 2007 Zürich survey: comparison of sample with Zürich city census*

Variable	Survey average	Population average	Difference [%]
Age	50.76	46.76	+8.5
	% in Survey	% in population	
Gender [male]	43.6	47.9	−4.3
Education			
State minimum	8.0	19.2	−11.2
Apprenticeship completed	31.8	31.3	−0.5
Upper secondary graduate	8.3	9.2	−0.9
Applied sciences graduate	20.8	10.7	+10.1
University graduate	26.0	17.1	+8.9
Car availability			
Always	44.6	42.8	+1.8
Sometimes	17.0	18.4	+1.4
Public transport pass			
Half-fare	49.5	37.9	+11.6
Regular pass	24.6	14.2	+10.4
Other passes	13.8	18.7	−4.9

Note: * Age, gender and education data from the 2000 Census (FSO 2000); transport data from 2005 Microcensus Travel (ARE and FSO 2007).

A first point of discussion concerns answer validation. Respondents were able to name a total of 17 relationships with the first name-generator and 32 additional non-overlapping relationships with the second, producing a possible total of 49 alters. In fact, the lists could have been extended if necessary. The range of named relationships is 1 to 49. The maximum number of reported relationships was reached once and the mean was 12.35 relationships. Compared to the possible number of 49 relationships, the exhaustion rate (mean relationship divided by the possible number of 49 relationships) is 25.2%, indicating that respondents had sufficient possibilities to cite relationships.

The accuracy of the named alter's home location is measured by the rate of complete street addresses compared to less accurate information such as street intersection, corner, landmark, etc. Even though the interviewees were asked to give information of their named alters as accurate as possible, not all alter's home locations can be identified down to the street-address. This has several reasons, but mainly because of lack of recall and privacy concern. The protocol tried to reduce the first cause by encouraging the respondents to prepare for the interview by having personal sources such as e-mail archives, telephone bills, address books, telephone numbers stored on mobile phones and Christmas mailing lists at hand during the interview. Further the respondents were encouraged to complete missing information after the interview if possible, and then sending in the completed survey. The privacy concern was addressed during recruiting, in the introduction letter sent with the questionnaire and the interview itself. It was ensured, that all the given information will only be used for anonymous statistical analysis, and that the address data would not be kept permanently electronically or physically. Nevertheless, about 12% of the interviewees refused to reveal the exact street-addresses, but gave the most exact geocodable description of the alter's home location, so that they felt still comfortable giving the information away. The corresponding shares of the information received for alter's home location is:

- Postal address: 68.6%
- The next crossing street on the street of the alter's home location: 8.3%
- Closest identifiable landmark: 16.1%
- Neighbourhood: 3.1%
- Village: 2.8%
- City 1.1%

A total of 3,972 locations were then geocoded. Overall, 3,614 (91%) were successfully automatically geocoded within eight runs of re-correcting addresses, which were not recognized. The remaining 9% had to be corrected to next higher level of available information, such as zip code or city.

Results: Number of Social Contacts

The following section presents the central empirical findings on mobility and social networks. First, the size of the ego-centred network (degree) is analysed. Following that, the analysis focuses on the spatial properties and arrangement of social relations. Then the influence of the mobility biographical variables on social network attributes is measured, followed by the analysis of contact frequencies in comparison with distance between members of social networks. Also the contact frequencies of different contact means is analysed and how they interact with each other.

On average, the respondents named 12.5 contacts. This number is within the range collected in other studies of egocentric networks, as can be seen in Table 3.3. The comparison shows that respondents in a questionnaire with only one name generator named fewer contacts (Zuma Methodological Experiment; see Pfenning and Pfenning (1987) and Mexico City Study; see Bernard et al. (1998)). The Fischer instrument collected data for significantly larger egocentric networks. The contact details of our survey come from two questions asking the respondent to generate names, which is reflected in the number of friendships specified. When explicitly asked about relationships maintained during leisure activities, a relatively high proportion of respondents mentioned not very close relationships.

Table 3.3 2007 Zürich survey: Sample size and mean number of relationships in ego-centric social networks from different studies

	NCCS[a]	ZUMA[b]	JV[c]	MC[d]	HAIFA[e]	Toronto[f]	Own survey
Number of respondents	1050	239	98	99	262	87	307
Ø Network Size	18.5	7.8	21.8	10.1	14.0	11.7	12.4

Notes: [a] North California Community Study; see Fischer (1982); [b] Zuma Methodological Experiment; see Pfenning and Pfenning (1987); [c] Jacksonville Study; see Bernard et al. (1998); [d] Mexico City Study; see Bernard et al. (1998); [e] Haifa Study; see Fischer und Shavits (2005); [f] Toronto Study; Hogan et al., 2007; Carrasco et al., 2008.
Source: adapted from Wolf (2004, 248).

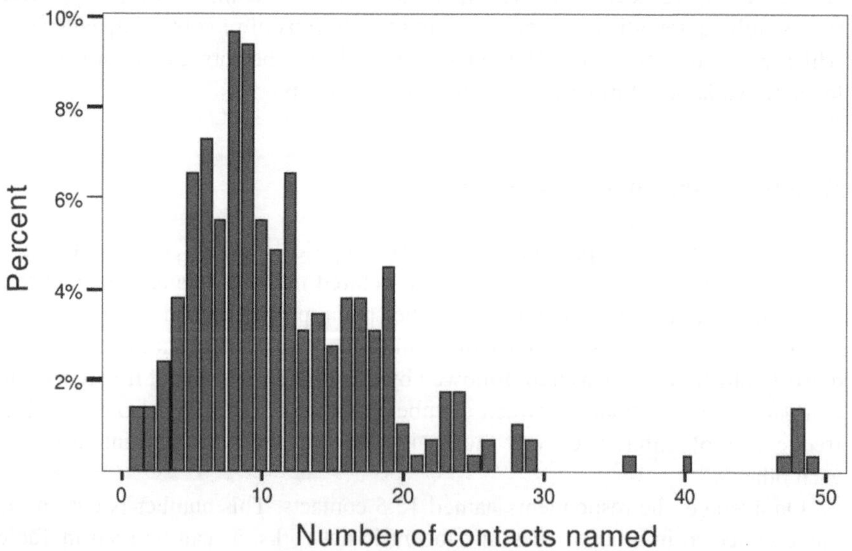

Figure 3.1a shows the distribution of the number of alters. The distribution is left skewed and has a variance of 73.0. The share of core relationships is 52% and drops, as expected, with an increasing number of reported relationships (Figure 3.1b).

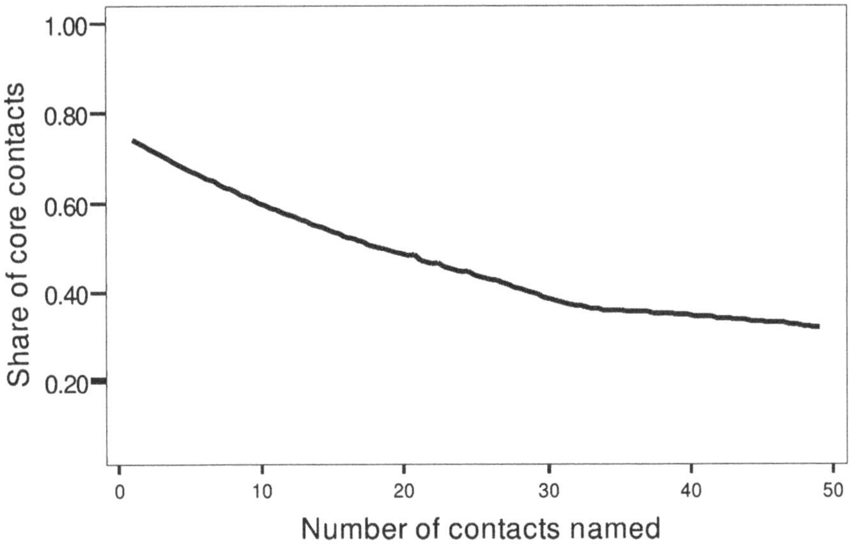

Figure 3.1 2007 Zürich survey: Distribution of the number of alters and their type of relationship a) Histogram of the number of alters named b) Share of core contacts vs. number of contacts named

Table 3.4 2007 Zürich survey: Parameter estimates for the negative binomial regression of the number of alters named

Variable	Mean	St. dev.	Parameter	Sign.
Constant			3.092	0.000
Age [years]	53.283	19.163	−0.040	0.002
Age^2/1000 [$years^2$/1000]	3.208	2.081	0.352	0.005
Annual or monthly public transport ticket [y/n]	0.853	0.893	0.242	0.042
Number of relocations [n]	5.963	3.116	0.038	0.003
University degree [y/n]	0.247	0.430	0.178	0.055
Part time employed [y/n]	0.170	0.382	−0.256	0.020
Retiree [y/n]	0.327	0.469	−0.302	0.045

Note: N = 305. Adjusted $R^2 = 0.13$.

To analyse the difference in the number of alters reported, sociodemographic, travel related, biographical and survey-specific dummy variables are employed in a negative binomial regression to account for the over-dispersed count outcome variable. After removing variables, which correlate highly with each other (limit = 0.5; e.g. working status and place of work), variables with a significance level lower than 0.05 were removed stepwise. The parameter estimates are reported in Table 3.4.

The question of cause and effect cannot be answered based on a multivariate analysis or the data available. The effects are considered to be reciprocal. This is especially true with regard to season ticket availability and use, as the availability and use of transport means has an influence on maintaining a social network and at the same time transport means usage is caused by participating in activities. Here is assumed that social relationships are the main driver for participating in activities. The age of the respondents shows a U-shaped influence. Younger people maintain many contacts and then the number declines with increasing age, whereas every additional year causes a lower decrease of the number of social relationships. The ownership of public annual or monthly transport ticket has a positive influence on the number of social relationships. Maintaining a bigger social network seems to be enabled by the ownership of mobility tools, but only the annual or monthly subscription to public transport tickets was highly significant. The number of relocations influences the number of relationships. The positive influence of the number of relocations indicates that people keep their important friendships after relocation even at a distance. By building up a social network at the new location and keeping in touch with old friends, numbers of social contacts increase. A higher education, at least a university degree, leads to a larger number of social contacts. But there is no clear trend visible between the number of alters and education. A rather high influence on the numbers of social contacts has the working status. In 27.3% of cases the original context of the acquaintance is work (41.0% friends, 25.9% family, 4.9% partner and 0.9% others) and therefore is the second most frequent original context. The big influence of the work status is therefore not surprising. Part time employees and retired people have fewer social relationships than full-time employees and equivalents e.g. students. It is surprising, that children have a positive influence on the amount of social contacts, because it would be expected that the additional workload for parents would decrease the number of contacts, but children open up the possibility to meet new possible contacts e.g. other parents with small children, parent-teacher conferences etc., which here outweighs the first effect.

Results: Spatial Distribution of Social Contacts

The most obvious way to measure the spatial dispersion of the egos' relationships is to measure the distance between their home locations. The addresses were georeferenced and the distances between the home addresses were calculated as great circle distances using equidistant cylindrical projection to account for the shape of the earth.

Distance Properties

The distribution of the great circle distances (Figure 3.2) between the respondents' residence and their relationships has three elements. Nearly two-thirds of the alters live locally within 25 km. The bulk of the remaining distances are divided into regional and national relationships (within 26–100 km 13%) and international relationships in Europe (within 101–1,000 km, 15%). A noticeable share of intercontinental links makes up the rest of 3% (Figure 3.2d). The peak at about 10,000 km marks the intercontinental distance between Zurich and the USA. The respondents mix the local/regional contacts of daily life with a multitude of non-local and often long distance contacts. Figure 3.2b shows that people are not only emotionally more closely attached to personally more important contacts, but also spatially closer.

The distance has a strong impact (note the log scale in Figure 3.2a) which is clearly visible in the histogram (Figure 3.2c), but the distribution does not follow a simple parametric distribution, which seems reasonable because people are not equally distributed over space because of places which are not habitable e.g. mountains or the sea and because people tend to cluster in cities.

Figure 3.3 shows a comparison between the shares of the population around the centroid of the respondents' residences near the main station of Zurich and the share of relationships in the same distance bands. The population shares were calculated using the Swiss census hectare raster data of the Federal Office for Statistics (FSO). This comparison is limited to Switzerland, because of missing comparable data for the surrounding countries, but this is not so critical as the population in Southern Germany, the nearest neighbour, which is at least 30 km away, is rather sparse. The decay of the relationships shares in the close-up range is remarkable in contrast to the population shares. Still, the share of contacts is over-proportional as visible in the very high ratios. The population share would be expected to grow with square of the distance for a uniform distribution. But it decreases in the close-up range slightly with the distance to the inner city of Zurich. From the distance band of 10 km on, the bigger cities of Switzerland and their agglomerations are clearly visible with the peak of Winterthur and Rapperswil and then Basle and Bern. The last two peaks are also slightly visible at the share of relationships. The ratio of these shares is independent of the distance bands and shows first a steep and then steady decrease with the distance from home. While people still select more than proportionally from those close-by, the range is today well beyond the distance of a 30 minute walk.

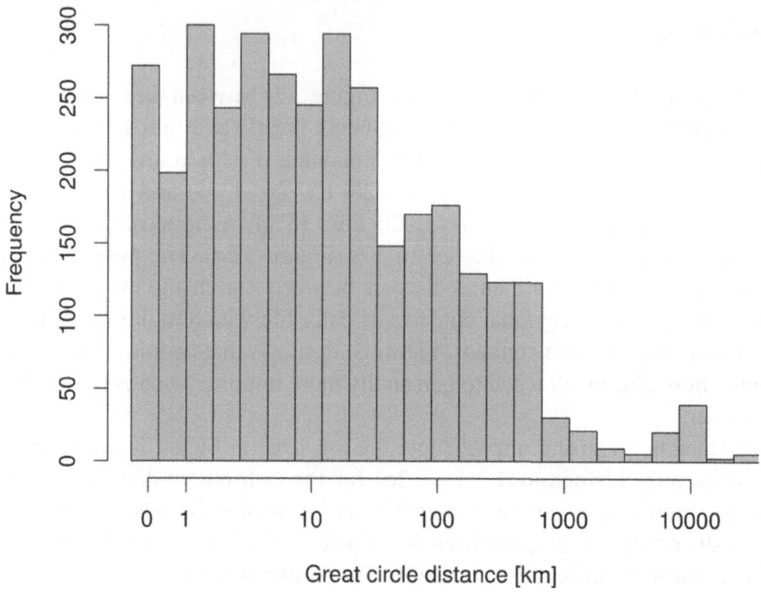
Frequency
0
50
100
150
200
250
300
0
1
10
100
1000
10000
Great circle distance [km]

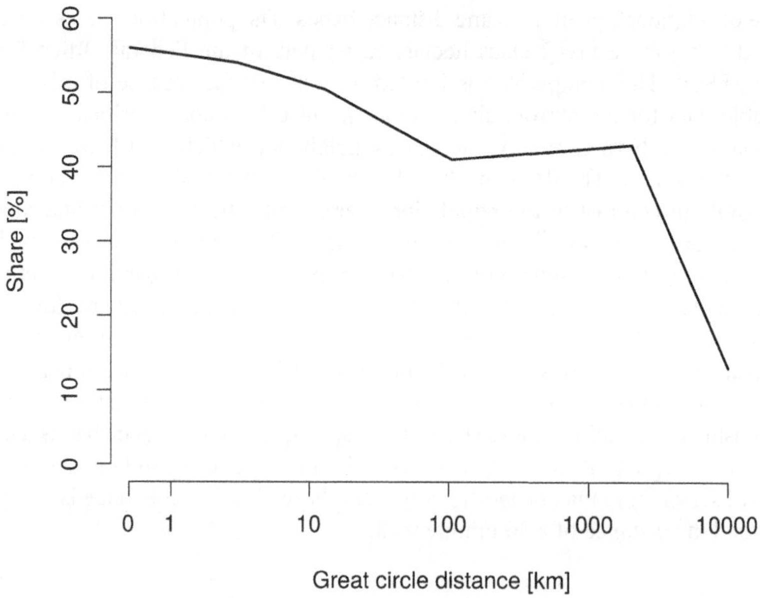
Share [%]
0
10
20
30
40
50
60
0
1
10
100
1000
10000
Great circle distance [km]

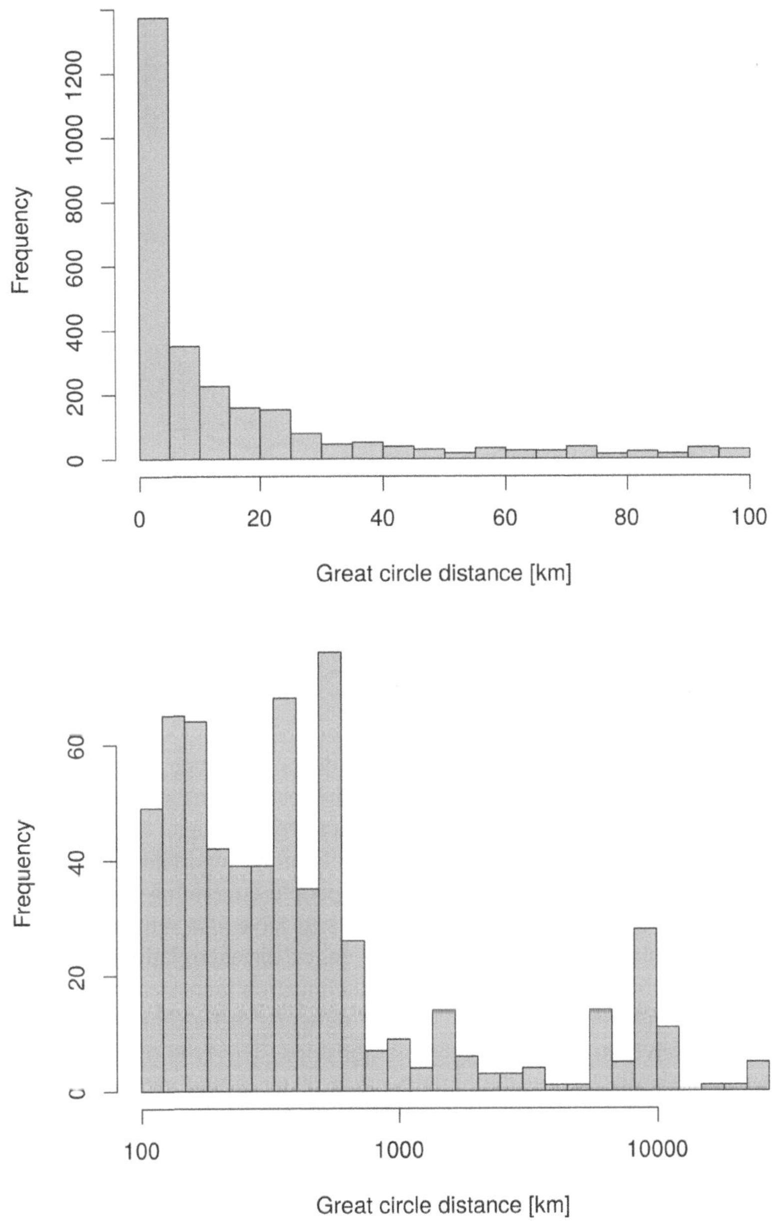

Figure 3.2 2007 Zürich survey: Distribution of the great circle distances between respondents and their contacts (note the different scaling of the x-axis) a) Over the whole distance spectrum b) Share of core contacts by distance c) Up until 100 km d) Over 100 km

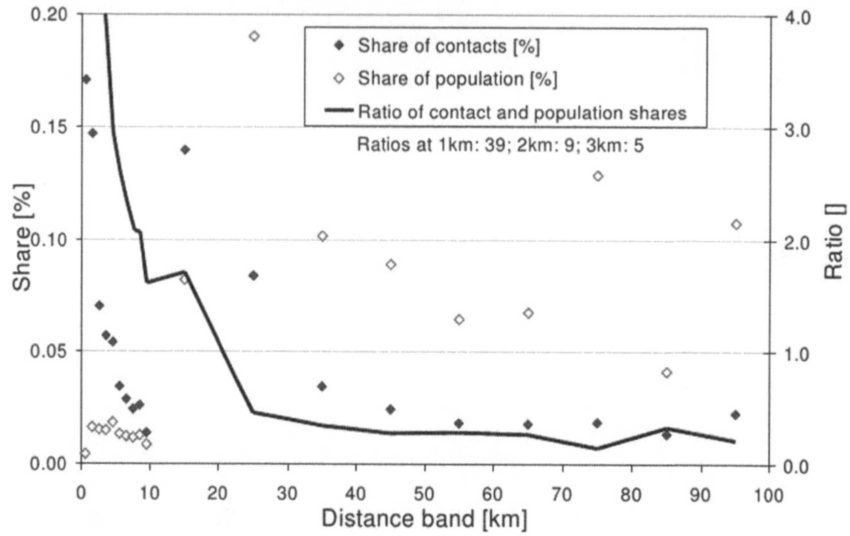

Figure 3.3 2007 Zürich survey: Contacts and population shares by distance band around Zürich

Note: Note the change in the width of the distances classes at 10km; due to lack of data the contribution of the populations in southern Germany had to be omitted.

The great circle distances between the respondents' residence and their alters' residences by socio-demographic attributes of the respondents are compared in Table 3.5. The socio-demographic differences are driven by age, education and their interactions plus the attachment to personally more important contacts.

Age makes a difference, with the younger people cultivating more distant relationships than the older ones. Gender seems to have also influence on the relationship distance, with men cultivating longer distant relationships. The education level indicates that a higher level of education increases the distance of relationships, and unlike with the social network degree, higher income people tend to have more long-distant relationships. The type of contact has an impact with family having the highest mean distance and partner with the shortest distances. This is expected, as we maintain contact with important people also at longer distances, but on the other hand want to be close to the most important people.

Table 3.5 2007 Zürich survey: Descriptive statistics of the great circle distances between home locations of alter and respondents [km] (with and without contacts within 25 km from the respondent)

Category of the respondent/ contact	Median		Mean		S.E. of the mean		Number of cases	
	with	w/o	with	w/o	with	w/o	with	w/o
Age								
Up to 30	13.3	92.8	316.2	707.0	65.9	165.4	675	239
30 to 40	9.5	166.5	580.8	1489.5	89.6	233.5	576	191
40 to 60	8.9	95.7	138.0	440.1	22.9	95.6	973	206
60 and older	8.4	121.9	219.5	773.9	43.9	169.0	1287	313
Sex								
Female	7.6	96.0	246.1	784.5	32.8	114.3	2045	523
Male	12.4	124.3	318.7	893.2	42.9	135.0	1466	426
Education								
N.A.	6.6	92.8	574.2	2751.3	243.4	1106.6	179	36
Obligatory schooling	4.3	75.8	210.6	826.3	113.8	458.8	254	62
Vocational training	8.4	86.5	130.4	523.0	32.4	140.9	1050	234
Highschool diploma	7.3	128.6	256.6	714.9	98.5	272.7	280	99
Further technical training	15.0	151.1	322.2	817.9	58.1	159.8	661	215
University degree	10.3	124.6	364.3	920.5	47.5	138.7	1087	303
Income [sFr/month]								
n.a.	13.3	156.6	264.8	733.2	83.1	236.6	375	117
0–1999	4.9	123.9	355.8	1341.8	81.2	295.2	462	121
2000–5999	8.5	108.7	233.3	679.8	31.2	99.1	1680	492
6000+	11.7	96.4	317.1	952.0	58.2	224.4	994	219
Type of contact								
Others	26.2	60.9	242.2	452.8	116.5	208.8	32	17
Friends	7.1	127.6	277.3	1030.1	43.6	172.8	1439	338
Partner	2.2	112.1	150.6	910.6	71.4	440.8	172	26
Family	21.0	96.6	304.7	637.6	45.8	106.8	911	357
Work mates	8.7	123.8	272.1	877.1	53.7	207.2	957	211
All	9.0	108.7	276.4	833.2	26.2	87.4	3511	949

To analyse the distances between ego's and alter's home location, the hierarchical structure of the data has to be considered. The datasets have an unbalanced hierarchical structure, resulting from respondents having different number of social contacts. The characteristics of ego-alter relationships cannot be treated as independent observations as they also depend on individual characteristics of the respondent. Therefore, multilevel linear regression modelling techniques are used to account for this structure, to estimate unbiased coefficients (for detailed information on multilevel modelling, see Snijders and Bosker 1999 and Goldstein 1995). The data for the distance analysis is structured in two levels: Level 1 includes information that depends on each ego-alter relationship, and Level 2 includes the egos socio-demographics as well as personal network characteristics, such as number of contacts and proportions of alters with common characteristics. Distances between egos and alters are skewed, resulting in outliers and extreme values. A residual maximum likelihood estimator (REML) is preferred over the least squares method (OLS) as OLS estimation is known for having disadvantages when dealing with skewed distributions including multiple outliers. Although both REML and Maximum Likelihood (ML) estimators fit parameters to the overall dataset, REML is preferred since it is more appropriate when estimating mixed models including fixed as well as random effects (Snijders and Bosker 1999). A logarithmic transformation was used to address the skewed distance distribution in the dataset. In order to avoid infinite values and loose observations with zero km distance in the transformation, all zero km distances were set to the minimum distance in the data. This procedure helps to avoid adding a constant to each distance, which has a relative higher influence on small distances compared to large ones.

The models include interactions and non-linear effects to represent the characteristics of the data as accurately as possible. Table 3.7 shows the parameter estimates, standard errors and t-values. After removing variables, which correlate highly with each other (limit = 0.5; e.g. education and income), variables with a significance level lower than 0.05 were removed stepwise.

Generally, it can be seen, that there is much more variance on Level 1 (Ego-Alter, Var[e] = 0.754) compared to the variance on Level 2 (Ego, Var[u] = 0.165), which means that the Level 1 variables can explain much more of the overall variance compared to the variables on Level 2. In general most of the people mix their local and non-local contacts, whereas there are only very few cases of respondents with either just local or just non-local contacts.

Regarding level 2 effects, distances decrease with age with a U-shaped influence (see the quadratic non-linear effect). Counter intuitive is the effect of the number of relocations, as one would expect, that people with many relocations maintain their previous contacts and thus maintain a more extensive network. The effect resulting from egos living with a partner in their households is strongly negative on distance. People living with their partner have smaller distances in their personal network, suggesting that those who care for others primarily maintain small distance relationships. Certainly, from an egos' perspective this effect can be related to the trade-off between the benefits from emotional contacts and distance costs, since nearby alters can be reached without spending much time and money on travel costs. Those savings in time and money can be invested into emotionally important relations, such as partnership or parenthood. The availability of internet increases the network distances. In general, the data show that people with internet resources maintain longer distance relations. In addition, higher education levels of the egos increases network distances. Considering that institutions for education are good places to meet others and establish relations, this result is not surprising.

Table 3.6 2007 Zürich survey: Multivariate multilevel analysis of the great circle distances between home locations of contact and respondents

Variable	Estimate	Std. error	t-value
Constant	2.878	0.536	5.372
Ego's characteristics (Level 2)			
Female	−0.138	0.064	−2.146
Age [years]	−0.023	0.011	−2.183
Age^2[years^2]	0.001	0.000	2.004
Number of relocations [−]	−0.121	0.057	−2.132
Living with partner [y/n]	−0.312	0.074	−4.212
Internet access [y/n]	0.237	0.077	3.081
Further technical training [y/n]	0.177	0.121	1.458
University degree [y/n]	0.324	0.133	2.436
Alter characteristics (Level 1)			
Immediate family [y/n]	−0.522	0.053	−9.85
Extended family [y/n]	2.321	0.932	2.491
Very close tie (emotionally) [y/n]	−0.256	0.052	−4.972
Known > 10 years [y/n]	−0.973	0.465	−2.094
Known > 10 years [y/n] Extended family [y/n]	−2.824	0.931	3.033
Known > 10 years [y/n] Number of relocations [−]	0.145	0.056	2.593

Note: Number of respondents = 265; Number of observations = 3,156; BIC = 8,547; Var[e] = 0.745, Var[u] = 0.165.

Regarding level 1, immediate relatives live closer to egos than friends, while extended family members live further away. This suggests that people stay in contact to their family members independently of geographic distances, which could be expected since family members are in general a reliable source of help and support, and may offer kinds of support that other social contacts may not. This is accentuated by the strong negative influence of important contacts on distance, as well as the negative influence on distance for contacts known longer than 10 years. This result also suggests, that contacts over longer distance might be maintained for a certain period of time, but the likelihood of maintaining such costly and high effort relationships over a longer period of time is lower compared to close distance relationships. There are two significant interaction effects with contacts known for more than ten years and extended family as well as number of relocations. The first interaction effect suggests that extended family relationships are over-proportionally maintained, when they are closer, and that respondents who move often maintain their long-time relationship also at a longer distance.

Social Network Geographies: Confidence Ellipses

The analysis above focused on the distance between the respondents' and their contacts' residence, but the distance alone ignores the pattern of the contact, e.g. the agglomeration of contacts, which cannot be measured just by distance and its distribution parameters. Biologists and more recently transport planners had to address the identical question of how to measure spatial distributions in their analysis of daily activity spaces. They proposed parametric, semi-parametric and non-parametric approaches to measure the size of the activity spaces (see Schönfelder 2010 for a review). The most popular, but also problematic approach is to calculate the size of the confidence ellipse, i.e. the two-dimensional generalization of the confidence interval. The Confidence ellipses are frequently used as an exploratory and hypothesis test-oriented method for the analysis of the relationship between two variables. They can be helpful to visualize and quantify the spatial distributions if the x- and y-coordinates are defined as locations of two (mutually dependent) variables. Confidence ellipses are analogous to the confidence interval of univariate distributions, where it defines the smallest possible (sub) region, in which a defined portion – 95% for a t-value of 1.96 – of the observations can be found. The calculation of the ellipses, however, is conditional on the assumption that the distribution follows a bivariate normal distribution. This is shown, at least approximately, for the distribution of activity locations (Moore 1970). But this does not necessarily hold for the spatial distribution of social relationships, which is not true as shown above. This leads to an overestimation of the calculated area. Also the symmetry of the confidence ellipse leads often to cases where half of the area covered by the ellipse is empty of locations and therefore is too big. Nevertheless, the confidence ellipses are here used as a descriptive representation of the spatiality of ego-centric social networks. As the confidence ellipses are analysed among each other, at least parts of the bias caused by the normal distribution assumption counts out.

The ellipsis is defined by the covariance-matrix of all home locations of one ego's alters, where the home locations are weighted by the frequency of visits. Figure 3.4 shows the graphical representation of a confidence ellipse for one respondent.

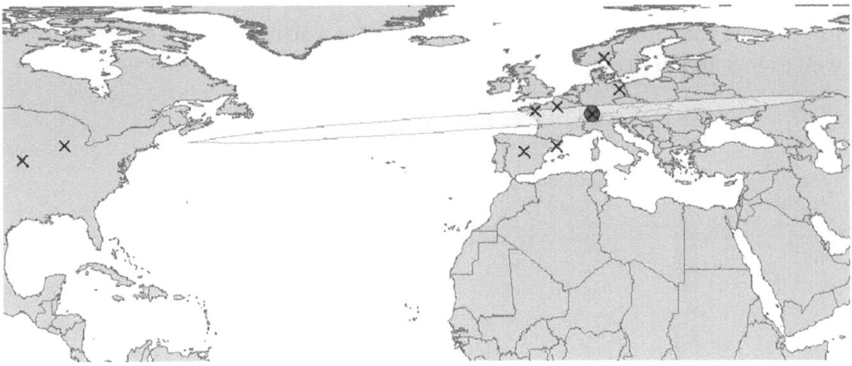

Figure 3.4 Example social network geography

Note: The respondent is female, 35 years old, full time employed and has moved 8 times in the last 22 years. The dark circled cross tags the current home location and the black crosses are the home locations of the alters.

Here, the 95% confidence ellipse is used, which by definition encompasses 95% of the geocoded points per ego. To avoid bias, the alters' residence locations are weighted by the annual frequency of face-to-face meetings. This has the purpose to represent the activity space in which the respondents' physical social world is most likely to take place. This also avoids giving too much emphasis on geographically distant contacts, which are not visited often. The areas are calculated using the Behrmann-projection, which is an equal-area cylindrical map projection. This is important as the calculated geographies vary strongly and are therefore prone to distortions.

The confidence ellipse measures geographical patterns with just three parameters (length of the main axis, ratio between the two axes and angle of the main axis). This makes it an easy to use instrument for the analysis of the spatial distribution of social relationships. For analysing the geographic patterns of social geographies we use the area of the confidence ellipse as a measure of the spatial expanse of a social network.

The distribution of the size of the 95% confidence ellipses seems to follow a log-normal distribution (Figure 3.5), if we ignore the third of the respondents who have a local set of contacts. The patterns of the socio-demographic differences follow generally the patterns of the distances to the contacts' home location above (see Table 3.7). The age of the ego shows a U-shaped effect, where the middle aged egos have the largest network geographies. However, 50% of young people below 29 years old have at least 20,793 km^2 large network geographies, which is compared to other age groups the largest median value. Male egos tend to maintain a slightly larger network geography comparing median and mean to females. Income and education-related differences show no clear effects. High and low income categories can have spatially more expansive network geographies, as well as low and high education categories.

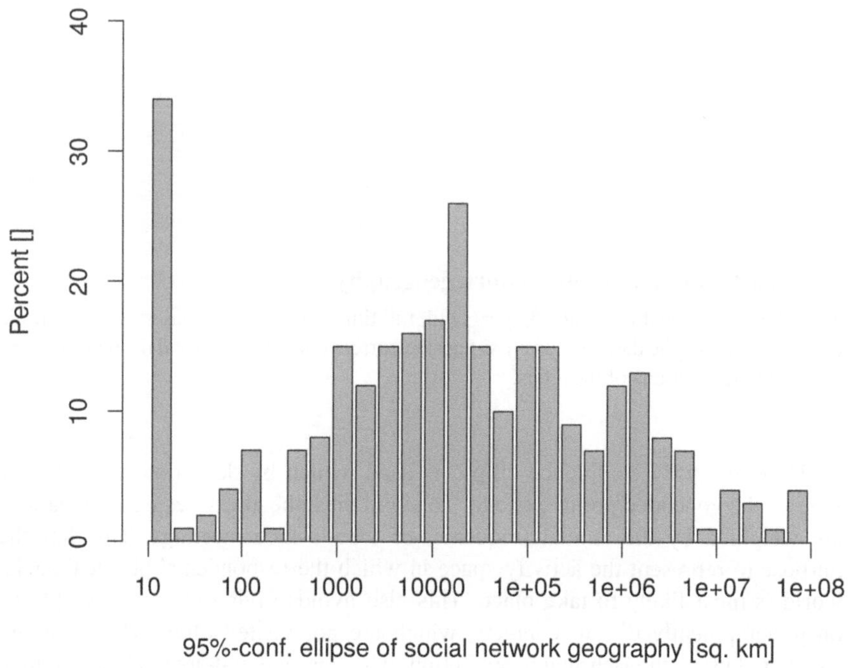

Figure 3.5 2007 Zürich survey: Distribution of the social network geometries

Note: Social Network geographies of less than 10 km² were coded as 10 km².

To investigate the differences in geographical size of social networks further, a multivariate analysis with the confidence ellipse as a dependent variable is employed. A model of the logarithm of the size of the 95% confidence ellipses as a dependent variable is considered as the values are all non-negative, with 33 zero value observations. The zero observations occur, when an Ego reported two or less alter's locations in their network. Conventional regression-methods, as the ordinary least square method, are not adequate for such censored values (Greene and Zhang 2003). A model, which is able to differentiate between limit-observations and non-limit-observations, is the Tobit Model. The Tobit model assumes that the limit outcome is determined by the level of the non-limit outcome. To test this assumption, a different model, which is also appropriate for the data can be compared to the Tobit model. This is Cragg's Model for Censored Data (Cragg 1971). It is used, when the assumption of the Tobit model, that the non-limit outcome is determined apart from the level of the non-limit outcome, is not true. Cragg's Model is a combination of the Probit model (for y=0) and the truncated regression (for y>0). The zeroes in our data have their origin in two different problems, the first problem is, that only around 70% of the geocodes have street address accuracy, while the rest has only zip-level, neighbourhood, or village/city accuracy, which leads to just one geocode for several contacts and the second problem is, that the confidence ellipse needs at least three spatially distinct locations to be calculated. The origin of the zeros in the data leads to the assumption, that the non-limit outcome is determined by the same level as the limit outcome, which is shown through the Probit and Tobit results (Table 3.8). The Tobit model was calculated after removing variables, which correlate highly with each other (limit=0.5), and variables with a significance level lower than 0.05 were removed stepwise.

Table 3.7 2007 Zürich survey: Descriptive statistics of 95% confidence ellipses of the social network geographies [10^5 km²] (with and without contacts within 25 km from the respondent)

Variable	Median		Mean		St. err. of the mean		Number of cases	
	with	w/o	with	w/o	with	w/o	with	w/o
Age								
Up to 30	0.27	0.27	66.73	68.29	40.37	41.29	44	43
30 to 40	0.12	0.18	36.19	41.76	29.71	34.25	49	39
40 to 60	0.14	0.25	17.74	20.8	8.38	9.78	78	58
60 and older	0.16	0.29	17.75	20.68	12.24	14.25	131	103
Sex								
Female	0.15	0.21	20.54	23.57	9.86	11.29	160	129
Male	0.21	0.31	37.68	42.64	17.83	20.14	142	114
Education								
N.A.	0.33	0.33	6.62	6.62	5.14	5.14	9	9
Obligatory schooling	0.04	0.14	1.71	2.14	1.18	1.46	28	20
Vocational training	0.06	0.15	5.74	7.03	2.56	3.12	101	71
High school diploma	0.16	0.19	51.37	55.65	35.09	37.94	27	24
Further technical training	0.21	0.34	35.51	39.39	25.97	28.78	64	55
University degree	0.38	0.40	55.04	59.34	28.93	31.14	73	64
Income [CHF/month]								
N.A.	0.54	0.93	4.86	5.46	1.63	1.82	50	40
0 to 1,999	0.14	0.18	52.74	57.39	25.56	27.7	41	34
2,000 to 5,999	0.15	0.20	22.34	24.55	13.38	14.69	132	111
6,000 +	0.09	0.22	41.18	51.83	26.89	33.76	79	58
All	0.17	0.25	28.53	32.52	9.83	11.18	302	243

Table 3.8 2007 Zürich survey: Parameter estimates for the Tobit regression of the logarithm of the size of the 95% confidence ellipses and the associated Probit model of the Cragg aproach Tobit model

			Tobit model		Probit model	
Variable	**Mean**	**St. dev.**	**Beta**	**Sign.**	**Beta**	**Sign.**
Constant	–	–	9.93	0.00	2.48	0.03
Age [years]	53.43	19.30	−0.29	0.00	−0.11	0.01
Age^2 / 1,000 [$years^2$/1,000]	3.226	2.09	2.95	0.00	1.09	0.01
Car ownership [y/n]	0.472	0.50	1.61	0.01	0.19	0.37
Number of relationships [−]	12.406	8.45	0.20	0.00	0.08	0.00
Education/workplace changes [−]	3.336	2.47	0.29	0.02	0.05	0.28
Further technical training [y/n]	0.213	0.41	2.48	0.00	0.58	0.04
University degree [y/n]	0.245	0.43	2.62	0.00	0.40	0.16
Income >6,000 CHF/month [y/n]	0.262	0.44	−1.64	0.03	−0.28	0.24
N				286		241
Goodness-of-fit	Adjusted R^2 = 0.25				chi² (8 df) = 47.31	

The analysis of the Tobit results shows that there are different factors, which influence the social network geographies. The first group consists of socio-demographic variables. The model results indicate that young people with a high education and with a low- or a middle income tend to maintain a more spatially distributed social network. The influence of the age and the education is similar to their influence on the numbers of relationships. The influence of the income seems to be unexpected as a spatially more distributed social network is expensive to maintain. An interpretation of the negative influence could be that a higher income is often linked to a higher workload and more responsibility, which leads to a higher value of time for these persons. As travel costs have decreased over the decades (see for example Fröhlich and Axhausen (2002) for Switzerland or Bruegman (2005) for the USA), the time costs seem to exceed the financial costs of travelling. The ownership of a car has a positive influence on the size of the social network geographies, even if the ownership of a car does not contribute to the maintenance of contacts over the distance (see above), it is an indicator of mobile people. The number of relationships has an influence as mentioned above, as it is correlated with the share of non-core contacts, as it is now possible to maintain spatially more widely distributed network of weaker ties with less frequent face-to-face contacts in combination with telecommunication contacts. The number of education or workplace moves is a biographical influence on the social network geographies. It seems that being less anchored in space and being professionally flexible have a positive influence on the size of the social network geographies, while surprisingly, the spatial distribution of the education and workplace changes, measured by their confidence ellipses, has no significant influence on them. Overall the model explains 25% of the variance of the social network geographies.

Results: The Influence of Mobility Biography

Expanding on the previous sections results, following are the relationships between mobility biographical events, socio demographics, distance and social network geography explored in more detail. The sociology of space claims that actors define social space by the spatial arrangement of those objects they find socially relevant (e.g. Löw 2001; Sommer 1996). Social relationships can mark the physical expansion of social space. Recent sociological time diagnoses have shown that spatial mobility is on the rise in modern societies (e.g. Adey 2013; Beck 2008; Castells 2002; Urry 2000 and 2008). The theoretical analysis of modern societies includes a discussion of the requirements for spatial flexibility at a micro-sociological level. This is often taken to refer to frequent changes in domicile for education and career, which, in turn, can lead to a spatial expansion of places of activity. The empirical section below will investigate whether the individual mobility biography extends the spatial arrangement of social relationships.

There are many events in the life cycle that may allow or even necessitate such forms of mobility. These events make up a person's 'mobility biography' (Scheiner 2007; Beige and Axhausen 2008). The relatively new concept of a mobility biography has experienced surprising attention in recent years, although its meaning and methodological ramifications continue to vary widely in the literature (Heinickel and Dienel 2006). The mobility biographical approaches in transport studies are focusing on the succession of primary residences, the phenomenon of secondary residences, jobs and how mobility strategies are shaped by a person's biography (Lanzendorf 2003; Prillwitz 2008). Building on this concept, the following hypotheses have been formulated for the empirical analysis.

Hypothesis 1: Key Events in the Housing, Educational and Employment History

This section will examine mobility biographies for their impact on the spatial arrangement of social networks. Of interest is the extent to which the social activity space can be explained by the number of key events in a person's housing, educational and employment history. This approach builds on the hypothesis that social reference points are located spatially by events in a person's mobility biography which also actively constructs that person's social space. A person's 'ordering of living beings and social goods' (Löw 2001) is based on the extent to which that person's social space was opened by his or her specific and unique biography. These lead to Hypothesis 1.1:

- Hypothesis 1.1: The more eventful a person's housing, educational and employment history, the more spatially dispersed the person's social network.

On the other hand, a spatially stable life, defined by Simmel (1908) as a spatial fixation of one's interests, leads to a spatially compact social network. On an individual level, this can be correlated with a person's having a fixed domicile: Nowicka (2007: 73) notes that 'fixation has [a] social dimension in which the focal point is a point of social interaction'. It can consequently be assumed that fixed social contacts in a given place can prevent a person from taking advantage of various opportunities to be mobile or from being felt forced to move. A strong attachment to a particular place of residence can be correlated with strong social integration and interaction with one's close neighbours. This leads to Hypothesis 1.2:

- Hypothesis 1.2: The more stable a person's housing history, the more spatially compact that person's social network will be.

Hypothesis 2: Choice of Transport Mode

Social relationships require mobility and generate travel. Maintaining contacts requires a means of transportation, whether a car or a public transport season ticket. The available transportation plays a key role in social integration, because it allows people to use aeroplanes, automobiles, bicycles, bus tickets, etc. to meet significant others. These means of transport allow spatial distances to be overcome, allowing for face-to-face contacts. Urry concludes that 'much social life could not be undertaken without the flexibilities of the car' (Urry 2000: 23).

For cultural sociologist Burkart (1994: 224), the car can even be considered the basic equipment of a full member of society. This suggests that the availability of a private vehicle increases social integration, because it makes it easier to travel to meet with friends, relatives and acquaintances. The literature also suggests that people with cars have a geographically wider social network (Canzler and Knie 1998: 121). On the other hand, those without a car would apparently have a less distance-intensive lifestyle (Canzler and Knie 1998: 97). Whether this is the case remains to be tested empirically. This leads to Hypothesis 2:

- Hypothesis 2: People with long-term access to a car find it easier to maintain spatially-remote contacts and therefore have spatially-dispersed social networks.

Hypothesis 3: Migration

Simmel (1908) states that spatial fixations can be marked by where social contacts reside. Migration research has shown some important correlations along these lines. For example, it can be shown that migrants reduce the risks of their target destination by choosing a place where an acquaintance or relative already resides (Grieco 1995). The places associated with social relationships are considered potential destinations for emigration (Beck 1997: 58). An important aspect is the interaction of migrants between the place of origin and the place of emigration, as seen in various research approaches (e.g., Espinosa and Massey 1997; Fall 1998; Beck-Gernsheim 2007). These show that social relationships create a strong attachment to places of origin even after emigration. People with a migration background do not simply leave the places of their origin behind, unless occasional visits are rendered impossible by war or political reasons. Otherwise, weddings, school holidays and other occasions are often reason to return to the place of origin. This leads to Hypothesis 3:

- Hypothesis 3: People with a migration background have spatially-dispersed social networks.

Hypothesis 4: Associations

Social communities can manifest material fixations in space. For example, for faith communities, this can be expressed by the house of worship that stands at the centre of its members' activities (Frisby 1984: 128). Also, membership in groups with shared interests, which is accompanied by social obligations and checks, can result in a spatial fixation. This then, according to the thesis, can result in a strong local social integration, which leads to the fourth hypothesis:

- Hypothesis 4: People in established local interest and social groups such as associations have spatially compact social networks.

The four hypotheses will be tested empirically in the following.

Structural Equation Models

Structural equation modelling (SEM) is used to test the hypotheses formulated here. Structural equation modelling is a structural test method for testing hypotheses (Backhaus et al. 2006: 334). Hypotheses were formulated in advance to allow SEM to be used to model the spatial distribution of egocentric networks as a function of mobility biographical events. The hypotheses function as an ex ante model.

SEM models can distinguish between a measurement level (manifest variables) and a structural level (latent constructs). As is the case with statistical model tests in general, this method cannot provide causality; the procedures only test whether a causal relationship is not rejected based on the hypotheses postulated and the underlying assumptions (Reinecke 2005: 12) . Whether a formulated model fits the underlying empirical data depends on how well the estimated covariance matrix corresponds to the empirical covariance matrix.

The model's quality was assessed with model quality parameters, such as the correlation between the chi square and the degrees of freedom (df). Thus the model can be evaluated for their goodness-of-fit to the observed data. The strength of the coefficient of regression, also path coefficients for manifest variables and factor loadings for latent constructs, is indicative of the degree of correlation. These are also tested for significant impact. Three main types of model testing must be distinguished:

- a strictly confirmatory approach using a model which confirms or rejects hypotheses
- a strictly confirmatory approach based on multiple models to test hypotheses and help researchers opt for a model
- the successive improvement of an initial model (to approximate the data structure) (Jöreskog und Sörbom 1993: 115).

The latter case is the one most commonly used in research (Reinecke 2005: 1). Therefore, it is important to note in the modelling phase that a confirmatory procedure often cannot be consistently performed. More commonly, adjustments to the model result in a back-and-forth between a model and the data. Even here, the final model is produced by adjusting the base model formulated on the basis of the stated hypotheses to the data structure.

Model Structure: Base Model

Table 3.9 shows the base model. Since mobility biographies as such cannot be measured, they are represented in the base model with a multi-dimensional latent construct. This requires operationalizing the mobility biography.

Hypothesis 1.1 postulates that a larger ego network geography is correlated to an increased number of events in that person's housing, education and employment history. To test this hypothesis, a factor describing the subject's mobility biography is generated by variables that measure the person's spatial flexibility over the course of his or her life (abbreviated here as: Mob-Bio [Migration]). This latent construct contains the person's number of stays abroad lasting longer than three months, the number of professional and residential relocations, the duration of his or her education measured in years, and area circumscribed by the person's places of education and employment to date (confidence ellipse).

While Hypothesis 1.1 addresses the flexibility in the subject's curriculum vitae, Hypothesis 1.2 considers the level of stability in his or her biography and its potential negative impact on the network geography. It is of interest whether a spatially compact housing history results in a spatially compact network geography. The 'lifetime stability' factor (abbreviated here as: Mob-Bio [Fixation]) is operationalized with the spatial extent of places of residence to date and the

Table 3.9 2007 Zürich survey: Base model variables plus hypotheses about the direct effects on network geography

Hypothesis	Manifest or latent variable			Descriptive statistic	Effect on network geography
1.1	Mobility biography 'migration' operationalized with		factor-charge		γ
	1	Number of stays abroad (> 3 months)	β	Avg.: 0.4	
	2	Number of occupational changes	β	Avg.: 3.4	
	3	Number of residential relocations	β	Avg.: 3.9	
	4	Years in education	β	Avg.: 12.9	
	5	Educational and workplace geography	β	Avg.[1]: 6,000 km^2	
1.2	Mobility biography 'fixation' operationalized with		factor-charge		-γ
	1	Geography of previous residences	-β	Avg.[1]: 3,000 km^2	
	2	Length of residence at current residence	β	Avg.: 2.2	
2	Personal vehicle history: Number of cars owned			Avg.: 0.3	γ
3	Migrant [Yes]			Proportion 19%	γ
4	Active in clubs (min. 12 meetings per year) [Yes]			Proportion 39%	-γ
	Sociodemographics				
		Gender (female = 1)		Proportion 56.4%	-γ
		Age < 29 years [Yes]		Proportion 15.0%	γ
		Income CHF 6,000 + [Yes]		Proportion 40.0%	γ
	Number of contacts mentioned			Avg.: 12.5	γ

number of years the subject had been residing at his or her current residence at the time the data was collected. Such an operationalization is necessary because not every change in residence is a change in region (or neighbourhood) and this effect can thus be controlled.

Manifest variables are then additionally modelled to these two latent constructs. One of these is information about the primary use of a passenger vehicle at certain stages of life. The respondents were asked to identify their primary means of transportation at different points in their biographies. The quotient of the years the subject has held a driver's license and the number of years when a personal vehicle was used as the main means of transportation is added to the base model as an index. This is used to test the hypothesis whether the use of a personal vehicle supports a broader spatial dispersion of social networks (Hypothesis 2).

In addition, a dichotomous variable is included with a value of one if the respondent has a migration background. A person was considered to have a migration background either if they were naturalized Swiss citizens or have permanent residency in Switzerland while holding a foreign passport. The leading hypothesis is that migrants remain in contact with the social network left behind in the land of origin and thus maintain networks that are more spatially dispersed (Hypothesis 3).

Another dummy variable investigated for its negative effect was for respondent participation in associations that meet at least twelve times a year. The hypothesis behind this is that people belonging to such clubs or associations have more spatially compact networks (Hypothesis 4).

Concerning socio-demographic and socio-economic variables, it is assumed that women maintain more compact network geographies than men due to the continuing strong division of gender roles with concomitant stronger ties to local activity spaces. Löw (2001) points out that the majority of girls are socialized to activity within a reduced spatial sphere, whereas boys learn to expand their sphere of activity (Löw 2001: 246 ff.) Similarly, it is hypothesized that incomes higher than CHF 6,000 per month and an age of up to 29 years can be connected with spatially extended networks, since these are traits that describe people who are more mobile and who have more mobile networks (Kesselring 2005).

The base model takes into account the non-geographic size of the subject's network by including the quantity of contacts. It is important to distinguish whether the spatial dimension of network geography increases with the number of nodes within the network or whether these are independent effects resulting from increased mobility. The key question then is whether a subject's mobility biography also has a significant effect on network geography, if the number of nodes is controlled for. This is an important consideration because each participant gave a different number of contacts. The advantage of the SEM method applied here is that both direct and indirect effects can be considered separately. Assuming that other variables remain the same (ceteris paribus), direct effects can be identified. The base model therefore tests the existence of a correlation between the spatial extent of the network and the number of contacts mentioned by the subjects. It can be assumed that a higher number of social contacts would increase the likelihood of their spatial dispersion.

Final Model

The base model has been successively adapted to the observations. The final model is shown in Figure 3.6 together with the path and regression coefficients. The model was gradually improved by removing insignificant influences and adding theoretically justifiable new paths that improved the model quality. It turns out that the two latent constructs migration and fixation have indeed developed while the measured variables have changed compared to the base model: variables with a factor loading less than 0.30 were removed in the first step because these were not sufficiently explained by the latent constructs. In a second step, these were then modelled as manifest variables. If there was no significant influence, they were then removed; if there was significant influence, they were kept. This explains why educational and professional geography and the years of education appear in the final model.

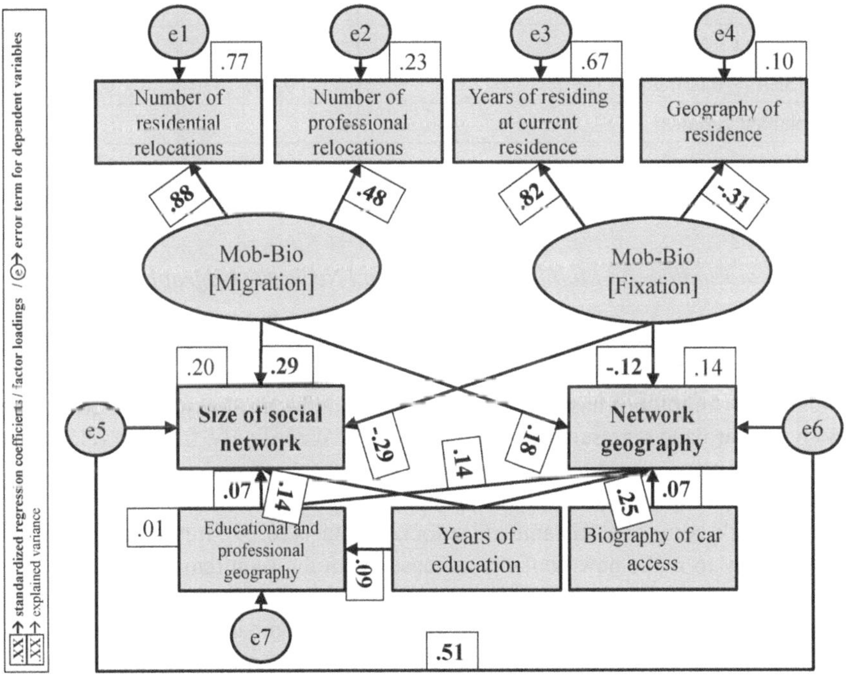

Figure 3.6 2007 Zürich survey: Final SEM Model of network geography size and of network size

Note: All paths shown significant at least at 0.1 level (detailed in Table 3.11).

Model Validity

The validity of the model is shown in Table 3.10. The final model fits the data structure well (GFI, AGFI, RMSEA, PC-Close and the ratio of chi^2 to degrees of freedom df). The AGFI indicator is only slightly below the acceptance criterion, such that it can still be deemed acceptable. Table 3.10 shows the different validity measurements. The final model can therefore be used for interpretation.

Table 3.10 2007 Zürich survey: Indicators of SEM model validity

	Base Model	Final Model
Sample size	298	298
Chi square	56.5	27.3
Degrees of freedom (df)	42	21
Chi square/df (< 2 = good; > 5 = not acceptable)	1.35	1.30
GFI (> 0.950 good)	0,903	0,964
AGFI (> 0.950 good)	0,848	0,945
PC-Close (> 0.050 good)	0,311	0,378
RMSEA (> 0.050 good)	0,061	0,048
Explained variance		
Network Geography	0.59	0.14

Empirical Findings on Mobility Biography and Network Geography

The significant and standardized regression coefficients confirm the suspected influence of the mobility biography on the network geography. Mobility biography (migration) has a significant positive influence on network geography with standardized regression coefficient of 0.18. Accordingly, frequent relocation, whether of place of residence or workplace, leads to a spatial expansion of social networks. The spatial flexibility in a person's biography is also correlated with the respondent's being able to name more social contacts (0.29). It is apparent that the opportunity to make new contacts increases with the requirement to deal with a new environment (Table 3.11).

The empirical results also show that a spatially compact housing history with few changes in residence has a negative effect on network geography, as can be seen in the mobility biography (fixation) (−0.12). This negative effect also appears in the number of social contacts mentioned (−0.29). It is likely that a stable housing history consolidates the size and spatial extent of the social network.

Table 3.11 2007 Zürich survey: Standardized direct effects on the network geography (final SEM model)

Hypotheses		Network Geography
1.1	Mobility biography 'migration'	.179**
1.1	Years in education	.252**
1.1	Educational and workplace geography	.144**
1.2	Mobility biography 'fixation'	-.122**
2	Personal vehicle history	.065*
3	Migrants [Yes]	n.s.
4	Active in clubs (min. 12 meetings per year) [Yes]	n.s.
	Sociodemographics	
	Gender (female = 1)	n.s.
	Age < 29 years [Yes]	n.s.
	Income CHF 6,000 + [Yes]	n.s.
	Number of contacts mentioned (covariance)	.511**

Notes: * significant to 0.1 level, ** significant to 0.05 level. *n.s.* = paths removed because not significant in base model.

The results of the model show that the use of a personal vehicle is slightly significant (90%-level) from a statistical perspective and has a minimal effect on the network geography (0.07). This result corroborates the results of the study of carless households by Preisendörfer and Rinn (2003). That study also showed that owning a personal vehicle had minimal to no influence on the number and distance of social contacts. This could be explained that both studies were conducted in cities, in this case Zürich. There are good public transport connections from Zürich, thus minimizing the effect of not owning a car on maintaining one's contacts. The postulated hypothesis could be more applicable to rural areas. Hypotheses 2 can only be confirmed with reservations due to the low levels of effect and the low confidence interval.

When comparing the influence of the number of contacts mentioned on the network geography, the final model shows a non-directional covariance between the error terms, indicating the best fit to the data structure (0.51). This means that the size of the network geography significantly increases with the number of contacts mentioned and, conversely, that the larger the area of the network geography, the more social contacts are named. It should, however, be noted here that the effects of mobility biographical variables are independent of the number of contacts mentioned (ceteris-paribus).

The number of years' education has a positive effect on the network geography (0.25) and on the number of contacts mentioned (0.14). It can be concluded that socio-spatial network building depends on education levels. Measured by the length of such education, this has a positive effect on the spatial extension of the network geography. A positive effect also exists between years of education and educational and vocational geography (0.09). The latter correlates positively with network geography (0.14) and the number of social contacts mentioned (0.07). In addition, the data structure shows that longer participation in educational institutions increases the chances of making contacts. People with longer periods of education named significantly more social contacts. In addition, longer periods of education results in a spatially dispersed array of schooling and job locations, which in turn leads to a spatial expansion of social contacts.

The final model shows that the two demographic characteristics of gender and income, as well as association activities and migration background did not show a significant influence and were therefore removed from the model. Accordingly, this study was unable to prove that associations lead to more local socio-spatial binding (Hypothesis 3) and that persons with a migration background maintain networks that are significantly spatially larger (Hypothesis 4).

The final model shows that events in a person's mobility biography, both fixation and migration, have a strong effect on the number of contacts mentioned and a significant direct effect on the spatial extent of networks. From these results, it can be concluded that breaks in one's biography lead to a spatial expansion of social networks, which supports Hypothesis 1.1. Conversely, spatial continuity in one's life, as measured in one's housing history, has a negative impact on the network geography, supporting Hypothesis 1.2. But it is also recognizable that people with low residential mobility mentioned significantly fewer social contacts in the survey.

Results: Staying in Contact: Modes and Frequencies

The data used in the previous sections does not only contain information about the home locations of social contacts, but also description of them. Especially, the original contexts of the acquaintance; and the frequency of communication by mode (face-to-face, phone, email, short text message (SMS)). Of interest here is whether the frequency of contacts can be explained, and in particular whether the distances between the respondents and their contacts play a role. A-priori, one would expect that distance will play a role if the effort involved with the interaction increases with distance. This should be true for:

- face-to-face interactions, which involve coordination efforts, travel time, travel costs, possible jet lag, accommodation costs, hospitality and gifts for the hosts visited;

- phone calls, which involve coordination efforts, time zone differences and fees, which in Switzerland increase with distance beyond the local call area and even more so beyond the national boundaries. At least there is a shadow price if the person has a prepaid budget of minutes under the contract signed with the telephone service provider. At the time of the survey, VOIP's market penetration was not yet high enough, and was not available yet for mobile phones to be a major factor;
- SMS messages, which involve flat fees, especially to recipients abroad, but no coordination costs. (The same shadow price argument holds here as well.)
- Email is essentially free at the point of use, either at home or at work, as flat rate schemes dominate and the contribution of email to any bandwidth charging scheme is negligible. There are no coordination costs.

Given the orders of magnitude of the cost differences between travel, calling and texting, one would expect that the strength of distance decay is different for these three modes. While this distance decay has been documented for travel since Lill's (1889) analysis of railway travel between Vienna and Prague in 1889 and as well as for telephone calls, for example in Rietveld and Janssen (1990), it has not been shown at the level of interactions between respondents and their contacts.

The rich literature on the interaction between travel and electronic communications has highlighted the potential for substitution and complementarity between the modes of contact, which necessitates a modelling approach that can account for these possible interactions.

Table 3.12 shows the characteristics of relationships in comparison with the type of relationship. This includes the comparison of the type of relationships with the number of contacts by contact means. Partners (4.9% of all) and those classified as 'other' (0.9%) are excluded from further analysis, as the former are special cases and the latter are rare. It has been shown that respondents overestimate rare events but assess high frequencies accurately (Schlich and Schönfelder 2001). The implied annual travel distances (the product of the great circle distance multiplied by the reported frequencies) does not confirm this point. These implied distances are comparable to the more careful estimates for annual leisure travel in the national travel surveys. Acknowledging that the national travel surveys often omit travel outside national boundaries and fail to capture journeys with mixed purposes, e.g., business travel mixed with a visit to relatives, the implied travelled distances come very close to the expected ones. In 2005 45% of the 13,000 annual kilometres were leisure travel in Switzerland, which is around 5,850 annual kilometres. 80% of leisure travel is declared visiting relatives and friends, which makes about 4,680 km of annual travelled distance per person socially motivated. If we consider, that leisure activity group sizes are around 2.3 on average (see Schlich and Axhausen 2003), the 11,000 implied travelled kilometres from the face-to-face contact frequency and distance between home locations becomes 4,783 kilometres, which is surprisingly comparable to the 4,680 km from the Swiss national travel survey.

If we add half of the estimated 6,000 km travelled outside the country (ARE and FSO 2007) (around 2,000 km for visiting and friends), we are even below the expected value. But we do not know who is visiting whom or to what extent people combine visits or see their contacts in groups during holidays at third locations. Accounting for all of this, the means are as expected. The much lower median highlights the importance of the rare, and probably overestimated frequencies of visits to contacts very far away.

Table 3.12 2007 Zürich survey: Characteristics of named contacts by type of relationship (excepting 'other' and partners)

		Relationship			
Variable	**Statistic**	**Friends**	**Family**	**Workmates**	**All**
Duration of relationship [years]	Mean	14	35	17	21
	Median	11	34	11	16
Face-to-face meetings per year	Mean	47	49	54	49
	Median	12	12	12	12
Telephone calls per year	Mean	27	48	35	35
	Median	12	24	12	12
Email written per year	Mean	10	9	22	13
	Median	0	0	0	0
SMS sent per year	Mean	12	16	24	16
	Median	0	0	0	0
Core contact	Share	47%	57%	55%	52%
Great circle distance [km]	Mean	277.84	313.37	272.10	286.10
	Median	7.21	20.49	8.70	9.95
Annual implied km travelled	Mean	7,816	16,712	7,675	10,669
	Median	1,918	1,466	1,063	1,325
Share		43.5%	27.5%	28.9%	100.0%

Note: N = 3302 for all variables except distance where N = 3168.

Figure 3.7 shows the shares of the modes as a function of the distance between the respondents and their contacts. The share of the face-to-face visits decreases quickly, as does the share of SMS, though rather more gently (statistically insignificant). Email gains share fastest with increasing distance because it stays constant in terms of frequency over the distance range. Phone calls gain share as well because they overtake all other modes in terms of frequency.

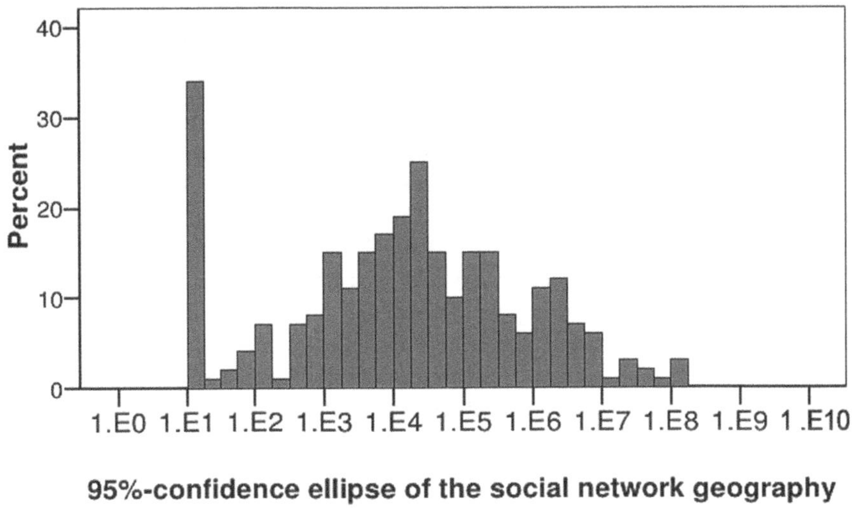

Figure 3.7 2007 Zürich survey: Share of the contact mode by log distance in deciles and zero between the respondent's and the contact's homes (for friends, relatives, workmates only)

Multivariate Analysis of the Number of Contacts in Social Network

To analyse the frequency of contacts between egos and alters, the hierarchical structure of the data has to be considered. The datasets have an unbalanced hierarchical structure, resulting from respondents having different social network degrees. The characteristics of ego-alter relationships cannot be treated as independent observations as they also depend on individual characteristics of the respondent. Therefore, multilevel linear regression modelling techniques are used to account for this structure, to estimate unbiased coefficients (for detailed information on multilevel modelling, see Snijders and Bosker 1999 and Goldstein 1995). Further are the contact frequencies not independent of each other, so we have to allow the model to let the different contact means correlate with each other. Approaching these data structures with an OLS estimator and an ordinary linear regression model would be a model misspecification. The results would be inexact parameter estimates with both unreliable standard errors and hypotheses tests. The assumption that all of the group structure is represented by the explanatory variables is unfounded. To capture these possibilities a three-level multivariate multilevel linear regression model is employed.

Because of the skewness of the contact frequencies between egos and alters, outliers and extreme values occur in the dataset. A residual maximum likelihood estimator (REML) is therefore preferred over the least squares method (OLS) as OLS estimation is known for having disadvantages when dealing with skewed distributions including multiple outliers. Although both REML and Maximum

Likelihood (ML) estimators fit parameters to the overall dataset, REML is preferred since is more appropriate when estimating mixed models including fixed as well as random effects (Snijders and Bosker 1999).

The following model was estimated on data from 307 interviewees (Level 3) and 3,302 reported relationships (Level 2). The dependent variable (Level 1) is the annual frequencies of face-to-face, phone, short message service and email contacts between egos and alters. The independent variables at Level 3 are characteristics of the ego and on Level 2 the relationships characteristics.

The frequency of contacts by mode is measured as reported counts of days with a social contact within a year. This makes the Poisson distribution (rather than the Normal) appropriate since the Poisson mean is >0 and the observation is count data. In this case, the logarithm of the response variable is linked to a linear function of explanatory variables. As the Poisson model assumes that the covariates have variance equal to the mean, we test the model for this assumption on under- or over-dispersion with Pearson residuals, where we check if residual deviance is much different compared to the degrees of freedom. This is with 0.01 strongly the case, so we have to deal with under-dispersion. To deal with this, there are different strategies. Quasi-Poisson methods could be used, but these are not recommended, as there are no likelihood functions available for them. To treat over-dispersion one can introduce an observation level random effect, which is the same as using a log-normal mixing distribution, where the log-link is used for the Poisson distribution and the marginal distribution of the observation-level effects are assumed to be normal and models the excess variation (Hinde 1982). One can unfortunately not model under-dispersion the same way, as variance cannot be negative. So instead of a quasi-Poisson model approach we use a variance stabilizing transformation for count data with less rare observations (Box, Hunter and Hunter 1978). This makes the model linear and therefore also easier for interpretation. As a rule of thumb, the results from a transformed linear model for count data have unbiased estimator if the mean number of observations is larger than 8 (Snijders and Bosker 1999). With a mean of 28, this criterion is met in this case. The variance stabilizing transformation should be used to reach normality and equal variance, neither of which hold in most cases perfectly. However, simple power transformations often improve the situation considerably. We use the log_10 transformation. To avoid zeros in the log transformation, the zero values are set to their corresponding minimum value in each contact mean. The following parameter estimates of the regression coefficient can be interpreted as following: For a one unit change in the predictor variable, the difference in the logs of expected counts is expected to change by the respective regression coefficient, given the other predictor variables in the model are held constant.

For modelling the frequency of contact we start with an empty model and include variables stepwise. First only fixed effects and random intercept models are considered, where first variables on Level 3 are included and then as well on Level 2. Further random slopes and also interaction effects within and between levels could be considered if significant and meaningful. The models are then compared by ANOVA to evaluate the justification of the additional variables.

In Table 3.13 all significant variables of the full model are shown. The overall model goodness of fit statistics are highly significant. The bulk of the explanatory power is due to the Level 2 differences. The intraclass correlation is the proportion of the variance that is accounted for by the group level compared to the total variance. There is more variance on the Level 2 than on the Level 3.The intraclass correlation on Level 3 is 0.39 for face-to-face, 0.32 for phone, 0.43 for Email and 0.56 for SMS contact frequencies. This means, that people in general are more homogeneous in their behaviour of how frequent they meet friends and relatives, but very heterogeneous among their different alters.

The correlation between the different contact modes are as expected, where especially phone and SMS contact frequencies are positively correlated with face-to-face contact frequencies, as they are used to coordinate meetings and are immediate means of contact. Email is negatively correlated with face-to-face contacts and seems to be rather complementary to face-to-face contacts; probably also because of its use for longer distances where time difference and language barriers are important. In any case, email is a different type of contact mode, much less personal and immediate. This is reflected in the low correlation between face-to-face contact and email contact frequencies.

Contact frequencies are decreasing significantly with an increasing network size. This confirms the idea of limited network capacities, as each social relationship needs time to be maintained (Dunbar 2003). In the models, only the variables, which are significant at least on the 0.1 level for one of the contact modes are kept.

All of the equations contribute significantly. The various modes of contact are complementary, except for E-mail. The socio-demographic effects are generally as expected: age is not significant for face-to-face contacts, but reduces the frequency of phone contacts for higher ages and especially for Email and SMS contact. This is expected, as older people did not grow up with the technology of these contact modes. Education does not show a clear picture and is not significant. High income increases the number of face-to-face interactions, while reducing phone contact. Regular car availability increases face-to-face contact, as it lowers the marginal cost of travel. The number of previous moves decreases the face-to-face contacts, as the spatial social network tends to be more scattered, as the number of moves increases. On the other hand the contacts with the phone increases slightly with the number of moves.

Table 3.14 shows the ANOVA between all the estimated models, which are presented in this section. All the models are nested within each other and are pairwise compared, so that each following column presents its model fit and the comparison statistics of the model in the column above. The empty model and the model with fixed effects on Level 3 are compared in the first two columns and so on. The model fit does improve from each model to the next. Whether the difference between two models is significant is tested with a likelihood-ratio test. All model improvements discussed above are significantly improving the overall fit of the model.

Table 3.13 Parameter estimates for full multivariate linear log-model with interactions

Contact Mode:	Face-to-Face			Phone			Email			SMS		
Parameter	*Est.*		*St.Err.*	*Est.*		*St.Err.*	*Est.*		*St.Err.*	*Est.*		*St.Err.*
Intercept	4.30	**	0.20	1.73	**	0.33	−0.46		0.32	1.58	**	0.35
Age_31–41	−0.17		0.17	−0.08		0.28	−0.64	*	0.28	−1.27	**	0.30
Age_42–58	−0.12		0.18	−0.59	*	0.30	−1.09	**	0.29	−1.95	**	0.32
Age_59–70	−0.21		0.19	−0.87	**	0.31	−1.48	**	0.30	−2.42	**	0.32
Age_71+	−0.20		0.18	−0.58	°	0.30	−2.17	**	0.29	−3.21	**	0.32
College	−0.05		0.15	0.30		0.24	0.03		0.23	−0.16		0.26
< 1999 sFr/month	0.06		0.18	−0.05		0.29	−0.35		0.28	0.13		0.32
> 6001 sFr/month	0.25	°	0.14	−0.34		0.23	−0.15		0.22	−0.25		0.25
Number of relocations	−0.04	°	0.02	−0.01		0.04	−0.05		0.03	−0.05		0.04
GA Yes	−0.10		0.14	0.30		0.23	0.45	*	0.22	0.06		0.25
Car Available Yes	0.21	°	0.12	0.26		0.20	0.08		0.19	−0.12		0.22
Degree	−0.02	*	0.01	−0.01		0.01	0.01		0.01	−0.04	**	0.01
Family Yes	0.29	*	0.12	1.23	**	0.19	0.83	**	0.19	0.72	**	0.18
ln distance	−0.38	**	0.02	−0.01		0.03	0.19	**	0.03	−0.05	°	0.03
Work Yes	−0.28	**	0.10	0.47	**	0.16	0.62	**	0.16	0.67	**	0.15
Relationship duration	−0.01	**	0.00	0.01	**	0.00	−0.01	**	0.00	0.00		0.00
Family*ln dist	0.03		0.03	−0.07		0.05	−0.18	**	0.05	−0.10	*	0.05
Work*ln dist	0.11	**	0.03	−0.01		0.05	0.04		0.05	−0.10	*	0.05
Signif. codes: 0: ** 0.01: * 0.05: ° 0.1: ‘ ‘ 1												
Loglikelihood	−26968.33											

Var. Comp.	Mode	Var.	Std. Dev.	Corr.		
Ego: Alter	FtoF	0.93	0.96			
	Phone	3.93	1.98	0.35		
	Email	3.93	1.98	0.09	0.16	
	SMS	3.33	1.83	0.25	0.20	0.24
Ego	FtoF	0.66	0.81			
	Phone	1.85	1.36	0.11		
	Email	1.66	1.29	−0.10	0.21	
	SMS	2.35	1.53	0.21	0.32	0.14

Table 3.14 ANOVA of all linear models

	Df	BIC	logLik	Chisq	Chi Df	P(>Chisq)	Pseudo R^2
Base model	25	59330	-29546	–	–	–	–
Only Level 3 variables	45	54567	-27071	4950	20	0.00	0.08
Full model w/o interaction effects	85	54778	-26988	167	40	0.00	0.09
Full model	93	54814	-26968	39	8	0.00	0.09

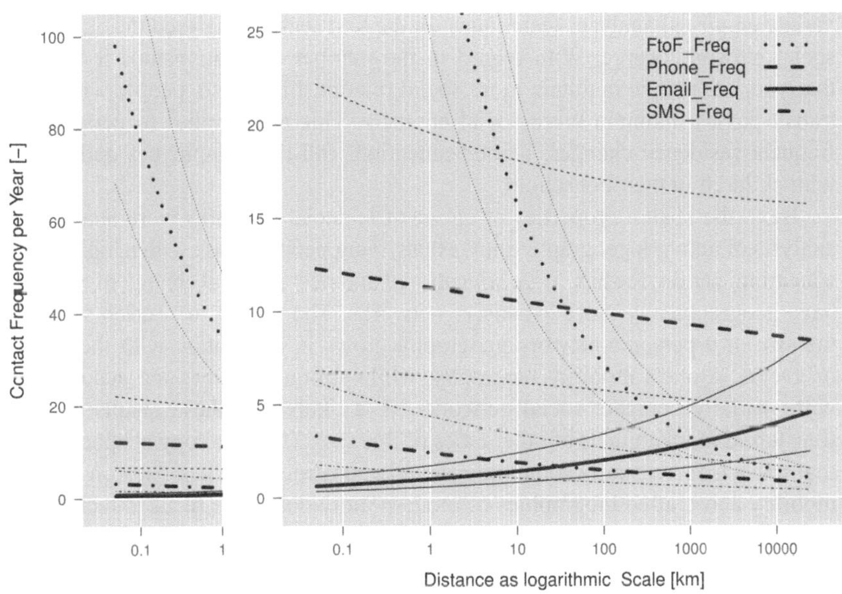

Figure 3.8 2007 Zürich survey: Predicted contact frequency per year and 95% confidence interval against ego-alter distance by mode

Note: Based on the model presented in Table 3.13. The left part of the Figure shows the contact frequency for ties within 1 km, whereas the whole distance spectrum is shown in the right part.

To visualize the effects of the model its predictions are shown against ego-alter distance for different contact means. The predicted values assume the fixed average values for the other contact means. The 95% confidence intervals are drawn based on the per-point variances. The predicted values are shown in Figure 3.8. It is clearly visible how strong the distance impact is on the face-to-face contact frequency, and the 95%-confidence band is very tight, which indicates the overall good model fit for face-to-face contacts. The distance decay for phone and SMS contact frequency is overall not very strong, and both decrease nearly at the same. Whereas the face-to-face contact is the dominant form in terms of frequency for shorter distances, the phone contacts become dominant at around 60 km. Note here also, how similar the model predicts the different curves compared to Figure 3.7. Email is the only contact form, which increases with distance in absolute terms. It surpasses face-to-face frequency at around 1,000 km, but stays behind phone contact frequency.

Synopsis and Further Research Questions

This chapter analysed three ego-centric network properties with uni- and multi-variate empirical analyses techniques. In the first section analysed the ego-centric social network degree. With regard to the number of relationships it is shown that younger people maintain more social relationships. Also, people with public transit passes maintain more social contacts. This also applies to persons with frequent residence changes, higher education, full-time employees and persons with children in their household.

Next, the spatial properties of ego-centric social networks were analysed. In analysis of network geography shows that young people, those with a high level of education, car ownership, more educational and occupational change of residence have geographically more expansive networks. The results for Zurich show that the size of a person's network geography grows in correlation with the number of events in one's mobility biography, while strong local ties are accompanied with spatially compact social networks. It is likely that these results could be transferred to inhabitants of other cities of this size. The sociology of space tells us that space is increasingly something actively constructed. When one considers the mobility and crafted biographies of modern individuals, the focus is increasingly on the spatial shaping of one's life, which, in turn, shapes the individual's social space. From a longer-term sociological perspective, there is much interest in the relationship between spatial and social structures, especially in connection with globalization, loss of borders, networking, digitalization and the frequently mentioned dissolution of national identities (Wellman 1996). The study showed that spatial flexibility leads to an expansion of the physical social space.

In terms of social capital, high levels of mobility are increasingly accompanied by problems with establishing stability and reliability. How is social cohesion possible in spite of high levels of spatial and social mobility in society? What is the impact of social and spatial mobility on the social differentiation of social contexts? How do people manage to maintain their network diaspora physically? Finally, one can conclude: the initial results presented here show that systematic differences exist in the geographical arrangement of social contacts, which in turn can be assigned to other social distinguishing traits, which is part of the task of research to follow. It also shows that these results can be used to generate social networks with systematic differences presented here. Such networks can then be used to simulate population wide influences from such networks. Such work is presented in Chapter 4.

Geographical distance influences how people interact today, as it did in the past (Mok and Wellman 2007) The effort involved in the different modes of communication (face-to-face, phone, email and SMS) matches the strength of distance decay in the frequency of those interactions. The patterns of the shares follow this differential distance decay. The frequency of face-to-face visits and their share falls quickest with distance.

While communication frequency via Email is unaffected by distance, its share rises fastest with an increasing distance. The reverse is true for SMS messaging, which stays stable in terms of share, while only slowly falling with distance in terms of frequency. The high frequency of phoning at short distances translates into growing shares, even if the absolute frequencies fall with distance. While other characteristics of the ego-alter dyad influence frequency and shares (duration of the relationship, context of the initial acquaintance), these are less influential in absolute terms than the socio-demographics of the person, especially income, age and the number of moves stand out among them. The impact of the characteristics of the dyad varies systematically with type of relationship, which reflect the underlying social expectations for relatives, friends and work mates.

Further it was shown, that the interaction of the different contact modes and Face-to-Face meetings are complementary.

Looking ahead, it would be useful at this point to bring in research on inequality, since differences in the appropriation of space could be socially critical, if such appropriation is assumed to reflect one's mobility biography (see Ohnmacht et al. 2009). For this purpose, the spatial world of one's life would have to be localized by representing network geographies. And as shown here further, the appropriation of space and maintaining that space trough social contacts is strongly correlated with the use of different contact means. The ability to maintain more extensive network geographies depends on acquiring the use of new telecommunication technologies. This could be again viewed in the context of inequality and enhance the notion of how socially critical the interrelation of network geography, mobility, mobility biography and the ability of using different contact means is.

References

Adey, P., Bissel, D., Hannam, K., Merriman, P. and Sheller, M. (eds) (2013), *The Routledge Handbook of Mobilities*, London: Routledge.

ARE and FSO (2007), *Mikrozensus zum Verkehrsverhalten 2005*, Berne and Neuenburg: Federal Office for Spatial Development and Federal Office for Statistics.

Axhausen, K.W., Frei A. and Ohnmacht, T. (2006), Networks, biographies and travel: First empirical and methodological results. In: IATBR (ed.), *The 11th International Conference on Travel Behaviour Research*, Kyoto: IATBR, http://term.kuciv.kyoto-u.ac.jp/iatbr06/

Backhaus, K., Erichson, B. and Plinke, W. (2006), *Multivariate Analysemethoden: Eine anwendungsorientierte Analysemethode*, Berlin: Springer.

Beck-Gernsheim, E. (2007), Transnational lives, transnational marriages: A review of the evidence from migrant communities in Europe, *Global Networks* 7(6): 271–88.

Beck, U. (1992), *Risk Society: Towards a New Modernity*, London: Sage.

Beck, U. (1997), *Was ist Globalisierung? Irrtümer des Globalismus – Antworten auf Globalisierung*, Frankfurt/Main: Suhrkamp.

Beck, U. (2008), Mobility and the cosmopolitan perspective. In: W. Canzler, V. Kaufmann and S. Kesselring (eds) *Tracing Mobilities: Towards a Cosmopolitan Perspective*, Aldershot: Ashgate, pp. 25–35.

Beck, U., Giddens, A. and Lash, S. (1994), *Reflexive Modernization: Politics, Tradition and Aesthetics in the Modern Social Order*, Stanford: Stanford University Press.

Beige, S. and Axhausen, K.W. (2008), Long-term and mid-term mobility decisions during the life course: Experiences with a retrospective survey, *IATSS Research* 32(2): 16–33.

Bernard, R.H., Killworth, P.D., McCarthy C, Shelley, G.A. and Robinsons, S. (1998), Comparison four different methods for measuring personal social networks, *Social Networks*, 3(12): 179–215.

Bonß, W., Kesselring, S. and Weiß, A. (2004), Society on the move: Mobilitätspioniere in der Zweiten Moderne. In: U. Beck and C. Lau (eds) *Entgrenzung und Entscheidung*, Frankfurt/Main: Suhrkamp, pp. 258–79.

Box, G. and Cox, D. (1964), An analysis of transformations (with discussion), *Journal of the Royal Statistical Society*, Series B, 26(211–52): 57.

Box, G., Hunter, J. and Hunter, W. (1978), *Statistics for Experimenters: An Introduction to Design, Data Analysis, and Model Building*, New York: Wiley.

Bruegman, R. (2005), *Urban sprawl: A Compact History*, Chicago: University of Chicago Press.

Burkart, G. (1994), Individuelle Mobilität und soziale Integration, *Soziale Welt* 45(1): 216–40.

Burt, R.S. (2000), Decay functions, *Social Networks* 22(1): 1–28.

Cairncross, F. (2001), *Death of Distance 2.0*, Boston: Harvard Business School Press.

Canzler, W. and Knie, A. (1998), *Möglichkeitsräume: Grundrisse einer modernen Mobilitäts und Verkehrspolitik*, Köln: Böhlau Verlag.

Castells, M. (2002), The space of flows. In: I. Susser (ed.) *The Castells Reader on Cities and Social Theory*, Oxford: Blackwells, pp. 314–66.

Cragg, J. (1971), Some statistical models for limited dependent variables with application to the demand for durable goods, *Econometrica: Journal of the Econometric Society* 39(5): 829–44.

Diaz-Bone, R. (1997), *Ego-zentrierte Netzwerkanalyse und familiale Beziehungssysteme*, Wiesbaden: Deutscher Universitätsverlag.

Dunbar, R.I.M. (2003), The social brain: Mind, language, and society in evolutionary perspective, *Annual Review of Anthropology* 32: 163–81.

Englisch, G. (2001), *Jobnomaden. Wie wir arbeiten, leben und lieben warden*, Frankfurt/Main: Campus.

Espinosa, K. and Massey, D.S. (1997), Undocumented migration and the quantity of social capital, *Soziale Welt* 12 (Special Issue): 141–62.

Fall, A. S. (1998), Migrants' long-distance relationships and social networks in Dakar, *Environment and Urbanization* 10(1): 135–45.

Fischer, C, S. and Shavits, Y. (2005), National differences in network density: Israel and the United States, *Social Networks* 17(2): 129–45.

Fischer, C. S. (1982), *To Dwell Among Friends. Personal Networks in Town and City*, Chicago: Chicago University Press.

Frei, A. and Axhausen, K.W. (2007), Size and structure of social network geographies, *Arbeitsbericht Verkehrs und Raumplanung, No. 439*, Zurich: Institute for Transport Planning and Systems, ETH Zurich.

Frisby, D. (1984), *Georg Simmel*, London: Tavistock.

Frohlich, P. and Axhausen, K.W. (2002), Development of car-based accessibility in Switzerland from 1950 through 2000: First results, *Arbeitsberichte Verkehrs- und Raumplanung* 111, IVT, ETH Zurich, Zurich.

FSO (2000), Volkszählung 2000, Neuenburg: Federal Office for Statistics.

Giddens, A. (1991), *The Consequences of Modernity*, Stanford: Stanford University Press.

Goldstein, H. (1995), *Multilevel Statistical Models*, Edward Arnold, London.

Greene, W. and Zhang, C. (2003), *Econometric Analysis*, Upper Saddle River, NJ: Prentice Hall.

Grieco, M. (1995), Transported lives: Urban social networks and labour circulation. In: S. Vertovec and A. Rogers (eds) *Explorations in Anthropology Series*, Oxford: Berg, pp. 36–50.

Heinickel, G. and Dienel, H.-L. (2006), Mobilitätsbiographien und Mobilitätsgenerationen. Ansätze für die Erklärung individueller und kollektiver Mobilitätskulturen. In: H.-L. Dienel (ed.), *Informationen zur modernen Stadtgeschichte. Themenschwerpunkt: Stadt und Verkehr*, Berlin: DIFU-Berichte, pp. 43–59.

Hinde, J. (1982), Compound poisson regression models, in proceedings of the *82th Internation Conference on Generalized Linear Models*, ed. R. Gilchrist, 109–121, New York: Springer.

Kesselring, S. (2005), Neue Formen des Mobilitätsmanagements. Mobilitätspioniere zwischen erster und zweiter Moderne, *Zeitschrift für Familienforschung*, 17(2): 129–43.

Kesselring, S. (2008), The mobile risk society. In: W. Canzler, V. Kaufmann and S. Kesselring (eds) *Tracing Mobilities: Towards a Cosmopolitan Perspective*, Aldershot: Ashgate, pp. 77–102.

Lanzendorf, M. (2003), Mobility biographies. A new perspective for understanding travel behaviour. In: IATBR (ed.) the *10th International Conference on Travel Behaviour Research* (Luzern: IATBR, http://www.ivt.baug.ethz.ch/allgemein/pdf/lanzendorf.pdf).

Ley, D. (2004), Transnational spaces and everyday lives, *Transactions of the Institute of British Geographers*, 29(2): 151–67.

Lill, E. (1889) Die Grundgesetze des Personenverkehrs, *Zeitschrift fur die Eisenbahnen und die Dampfschiffahrt*, 2(35): 697–706.

Löw, M. (2001), *Raumsoziologie*, Frankfurt/Main: Suhrkamp.

Manderscheid, K. and Bergman, M. (2008), Spatial patterns and social inequality in Switzerland – modern or postmodern? In: G. Pflieger, L. Pattaroni, C. Jemelin and V. Kaufmann (eds), *The Social Fabric of the Networked City*, London: Routledge, pp. 41–65.

Mok, D. and Berry, W. (2007), Did distance matter before the internet? Interpersonal contact and support in the 1970s, *Social Networks*, 29(3): 430–61.

Moore, E. G. (1970), Some spatial properties of urban contact fields, *Geographical Analysis*, 2(4): 376–86.

Nowicka, M. (2007), Mobile locations: Construction of home in a group of mobile transnational professionals, *Global Networks*, 7(1): 69–86.

Ohnmacht, T. (2006), Mapping social networks in time and space, *Arbeitsbericht Verkehrs und Raumplanung, No. 341*, Zurich: Institute for Transport Planning and Systems, ETH Zurich.

Ohnmacht, T., Maksim, H. and Bergman, M.M. (eds) (2009), *Mobilities and Inequality*, Aldershot: Ashgate.

Pfenning, A. and Pfenning, U. (1987), Egozentrierte Netzwerke. Verschiedene Instrumente – verschiedene Ergebnisse, *Zuma-Nachrichten*, 1: 64–77.

Pongratz, H.J. and Voß, G. (2003), *Arbeitskraftunternehmer. Erwerbsorientierungen in entgrenzten Arbeitsformen*, Berlin: Edition Sigma.

Preisendörfer, P. and Rinn, M. (2003), *Haushalte ohne Auto: Eine empirische Untersuchung zum Sozialprofil, zur Lebenslage und zur Mobilität autofreier Haushalte*, Opladen: Leske und Budrich.

Prillwitz, J. (2008), *Der Einfluss von Schlüsselereignissen im Lebenslauf auf das Verkehrshandeln unter besonderer Berücksichtigung von Wohnumzügen*, Leipzig: Helmholtz-Zentrum für Umweltfoschung UFZ.

Rai, R.K., Balmer, M., Rieser, M., Vikrant, V.S., Schönfelder, S. and Axhausen, K.W. (2007), Capturing human activity spaces: New geometries, *Transportation Research Record*, 2021: 70–80.

Reinecke, J. (2005), *Strukturgleichungsmodelle in den Sozialwissenschaften*, Berlin: Oldenbourg.

Rietveld, P. and L. Janssen (1990), Telephone calls and communication barriers, *The Annals of Regional Science* 24(4): 307–18.

Scheiner, J. (2007), Mobility biographies: Elements of a biographical theory of travel demand, *Erdkunde* 61(2): 161–73.

Schiffmann, S. (2005), *Mobilitätsbiographie:* Verkehrserzeugung im Alltag, Master's Thesis, MAS Spatial Planning, ETH Zürich.

Schlich, R. and K. W. Axhausen (2003), Habitual travel behaviour: Evidence from a six-week travel diary, *Transportation* 30(1): 13–36.

Schlich, R. and Schonfelder, S. (2001), Zwei Sichten der Aktivitatenteilnahme: Tagebucher und Aktivitatenhaufigkeitsbogen, *Stadt Region Land*, 71: 147–64.

Schönfelder, S. (2010), *Urban Rhythms: Modelling the Rhythms of Individual Travel Behaviour*, Farnham: Ashgate.

Schönfelder, S. and Axhausen, K.W. (2003), Activity spaces: Measures of social exclusion?, *Transportation Policy* 10(4): 273–86.

Sennet, R. (1998), *Der flexible Mensch: Die Kultur des neuen Kapitalismus*, Berlin: Berlin Verlag.

Simmel, G. (1908), Der Raum und die räumliche Ordnung der Gesellschaft. In: G. Simmel (ed.) *Soziologie: Untersuchungen* über *die Formen der Vergesellschaftung*, Berlin: Duncker und Humblot Verlag, pp. 460–526.

Snijders, T.A. and Bosker, R.J. (1999), *Multilevel Analysis. An Introduction to Basic and Advanced Multilevel Modelling*, London: Sage.

Sommer, R. (1996), *Personal Space*, Englewood Cliffs, NJ: Prentice Hall.

Spieckermann, H. (2005), Konstruktion sozialer Räume durch Netzwerke. In: M. Riege and H. Schubert (eds), *Sozialraumanalyse: Grundlagen – Methoden – Praxis*, Wiesbaden: VS Verlag, pp. 313–27.

Urry, J. (2000), *Sociology beyond Societies: Mobilities for the Twenty-First Century*, New York: Routledge.

Urry, J. (2008), *Mobilities*, Oxford: Blackwell.

Wellman, B. (1996), Are personal communities local? A dumptarian reconsideration, *Social Networks* 18(4): 347–54.

Wolf, C. (2004), Egozentrierte Netzwerke. Erhebungsverfahren und Datenqualität, *Kölner Zeitschrift für Soziologie und Sozialpsychologie*, 44 (Special Issue): 244–73.

Zinn, J.O. (2006), Biografische (Un-)Sicherheit in der Moderne: Zum Wandel von Selbstbindung und Widerständigkeit in Alltagstheorien und biografischer Forschung, *Sozialer Sinn* 7(2): 301–26.

Chapter 4
Personal Networks in a Population-Wide Network Structure

Matthias Kowald

Investigating the influence of social contacts on leisure travel patterns was the aim of the study by Frei and Axhausen (see Chapter 3) and numerous other survey projects in transport planning in recent years. Many of these studies used social network analysis (SNA) methods. They identified leisure travel as being primarily undertaken to join others in leisure activities; it is also referred to as 'social' or 'activity' travel. In focusing mainly on network topology statistics (i.e., number of personal contacts, geographical distances between tied persons, and contact modes and frequencies used to maintain relationships), these projects confirmed SNA-methods productivity, producing new empirical insights and results (Larsen et al. 2006; Carrasco et al. 2008; van den Berg et al. 2009; also see Chapter 3). However, one issue was ignored by previous studies: topology of a population-wide '*global*' leisure network.

Focussing on personal networks means that respondents' personal networks (usually) do not overlap in terms of shared social contacts. Surveyed network structures are isolated components of the global social network. However, in social reality people are connected to many others by direct, but to most others by chains of indirect relationships. In other words: People have a limited number of relatives, friends and acquaintances but when also accounting for friends of friends, or even higher orders of interconnectivity, they are linked to many others. Their personal networks are embedded in a population-wide network structure. This underlying network allows feelings (Christakis and Fowler 2009), styles and behaviours (Gladwell 2002), as well as contagious diseases (Mossong et al. 2008; Smieszek et al. 2011) to spread within a given population. Information on this population-wide network is needed to allow analyses of its structure and an implementation of a '*global*' leisure network in agent-based travel demand simulations.

The Institute for Transport Planning and Systems (IVT) of ETH Zurich and the Institute for Sea- and Land-Transport (ILS) of TU Berlin conducted a joint survey study between January 2009 and March 2011 to investigate this population-wide leisure network topology. The survey is presented in detail here. Starting with an introduction of the survey design, the chosen methodology to collect information on personal networks as well as a population-wide network structure is presented. This is followed by a discussion on the survey instrument. Survey protocol, closely linked to bias-decreasing measures developed for this project, is presented next. Accompanied by an analysis of people's response behaviour a fieldwork report is provided. It also includes information on relevant ethical and data protection issues. Discussing the fit between survey and target population and providing information on non-response behaviour aims to analyse the data in terms of data quality (most of this work was previously published in Kowald and Axhausen 2014). Finally, selected results from the survey are presented. This includes, on the one side, information on personal networks. However, instead of repeating the analysis from Chapter 3, additional aspects of personal networks are analysed, including issues of homophily and tie strength. On the other side, results about the population-wide network structure are presented. This includes general information on the overall network topology as well as an analysis of geodesic distances between the people involved.

Connected Personal Networks: Survey Methodology

The term '*social networks analysis*' does not refer to one fixed method of collecting and analysing data, but to a whole set of methods; an appropriate choice must be made according to project goals (Wasserman and Faust 2007). Basically, SNA-methods allow a focus on interacting actors, e.g. people, companies, or countries, and their ties at the same time; a tie is defined by whatever is circulating in the network, e.g. information, money, or simply willingness to spend free time together. So, explanations can consider actors' characteristics and their social environment. One sees how social structures, emerging from interactions, provide both opportunities and constraints for actors' activities. For our survey's aims, tools to collect two types of information were necessary: first, data on peoples' individual leisure network to replicate and enhance analyses from former studies and allow for comparisons and second, data on these individual network structures' embedded nature in a population-leisure relations network (most of this work was already published in Kowald and Axhausen, 2014).

Social relationships cannot be sampled per se; SNA-methods suggest using samples of actors for observation or questions about their relationships (Knoke and Kuklinski 1982). Leisure networks are normally developed over time, influenced by decisions about retaining or neglecting contacts. These structures are individual; people can move physically and socially throughout their lives, relatively unfettered by institutional settings like companies, municipalities, or even national borders. Given these factors, the '*personal network approach*' was used to survey leisure networks. It focuses on social environment, here in terms of leisure contacts, and allows a survey of information about representative population samples, which facilitates generalization of survey results. The counterpart to surveying personal networks is the '*full network approach*', where all actors and relations in an institutional setting are observed. This approach results in a comprehensive network structure picture and is only feasible for small populations (Marsden 1990).

When aiming for a population-wide network, the personal network approach has a serious disadvantage. Usually, studies on personal network structures seeking valid inferences recruit respondents from random samples. Following Groves (2004) participants (the survey population) are, therefore:

1. A subset of the frame population, including all people in the initial random sample from which they are recruited
2. An even smaller subset of the target population, defined as all individuals studied for whom inferences should be valid.

This size difference between survey and target population and smaller personal leisure networks, – averaging around 15 contacts as former studies showed (Larsen et al. 2006; Carrasco et al. 2008; Dunbar 2003; McPherson et al. 2006; Roberts et al. 2009; Silvis et al. 2006; van den Berg et al. 2009; also see Chapter 3) – makes overlap between personal network structures (two respondents either mentioning each other as social contacts or sharing a social contact) unlikely. In fact, no previous transport planning studies observed such overlaps; they all had to look at isolated personal network structures (for an illustration see Figure 4.1a).

A picture of a population-wide leisure network can be obtained by employing one of two survey methodologies. First, '*saturation sampling*' can be used, meaning all individuals in the relevant social structure are interviewed about their social relationships. Similar to the full network approach, this is feasible only for small target populations. The alternative is employing a link traced sampling methodology to survey information on personal networks and connections between those structures (Coleman 1958; Gabler 1992). Although there are several '*chain methods*' (for a general discussion see Erickson (1979)), they all ask some initial respondents for their social contacts and then use these contacts to continue recruitment. Depending on the method chosen, certain statistical features of the overall network structure can be calculated. The best-known chain methods are '*snowball sampling*' and the '*small world method*'.

Small world experiments use messages, e.g. letters or e-mails, that must reach predefined recipients (Milgram 1967; Kochen 1989, Schnettler 2009a). Participants are asked to either pass the message to the recipient directly, or to a person who would be able to forward it to the recipient. So, the experiment asks participants to use one of their social contacts to forward a message. This allows for estimation of average path length between two randomly chosen individuals within a population. One problem with this approach can be actors' limited information on their social environment. Because they lack knowledge about the overall network structure and possible shortcuts, chain lengths are often overestimated (Kleinberg 2000). Another problem results from questions on how to deal with disrupted contact chains that did not reach their destination (Schnettler 2009a; for an illustration see Figure 4.1b).

In comparison to a small world experiment, snowball sampling asks initial respondents, called '*seeds*', to report their social contacts. These social contacts are then again asked to mention their social contacts, a procedure which can be repeated for a predefined number of iterations. With the exception of the seeds all respondents in a snowball sample are mentioned by former respondents. The respondents on one iteration level introduce the frame population of the following iteration level. Finally, each seed makes up a component of directly or indirectly connected personal networks. Depending on the relation between survey and target population, overlaps between components are rare, a problem similar to surveying isolated personal networks employing the personal network approach without snowball sampling (see Figure 4.1c; for a schema of a snowball component with four iteration levels, see Figure 4.1d).

To summarize, snowball sampling benefits from the personal network approach – surveying information on personal networks in a large population – by avoiding exclusive focus on isolated network structures. Because only a few seeds are needed to find other members of a population, snowball sampling and other forms of link traced sampling methods can also be used to survey information on hidden or hard-to-reach populations. To collect data on connected actors, sociological (Schweizer 1998; Salentin 1999), anthropological (Jones 2003), and medical studies (Mathews et al. 2001) use snowball studies.

However, one must be aware that snowball samples include three major sources of bias. Initially, snowball sampling barely meets randomness criteria. Former respondents select all persons on the iteration levels, unlike the seeds, who can be recruited from a representative random sample. This '*selection bias*' results in unequally distributed probabilities for people to become part of the sample. Increasing response rate and surveying respondents' social contacts as completely as possible can help minimize selection bias. Although social contacts are still mentioned selectively, a complete personal network structure and high response rate can help to survey a valid picture of the variability of personal network structures. To reach this goal, respondents must feel confident about participating and mentioning their leisure associates. Furthermore, in snowball sampling, control should be maintained over snowball chains (when recruiting social contacts) in researchers' hands after experience with a study where respondents themselves were asked to recruit their social contacts, resulting in serious problems (Silvis et al. 2006; for similar designs where respondents are asked to recruit their contacts see Heckathorn 1997).

The second kind of bias deals with the well-documented phenomenon of homophily, describing a similarity between directly connected persons, whether in socio-demographics or in behavioral patterns (McPherson et al. 2001). '*Homophily bias*' poses a challenge for the survey methodology; the sample should ultimately represent the socio-demographical heterogeneity of its target population. However, when directly connected persons are similar to each other, seeds should be selected carefully to reproduce target population variability.

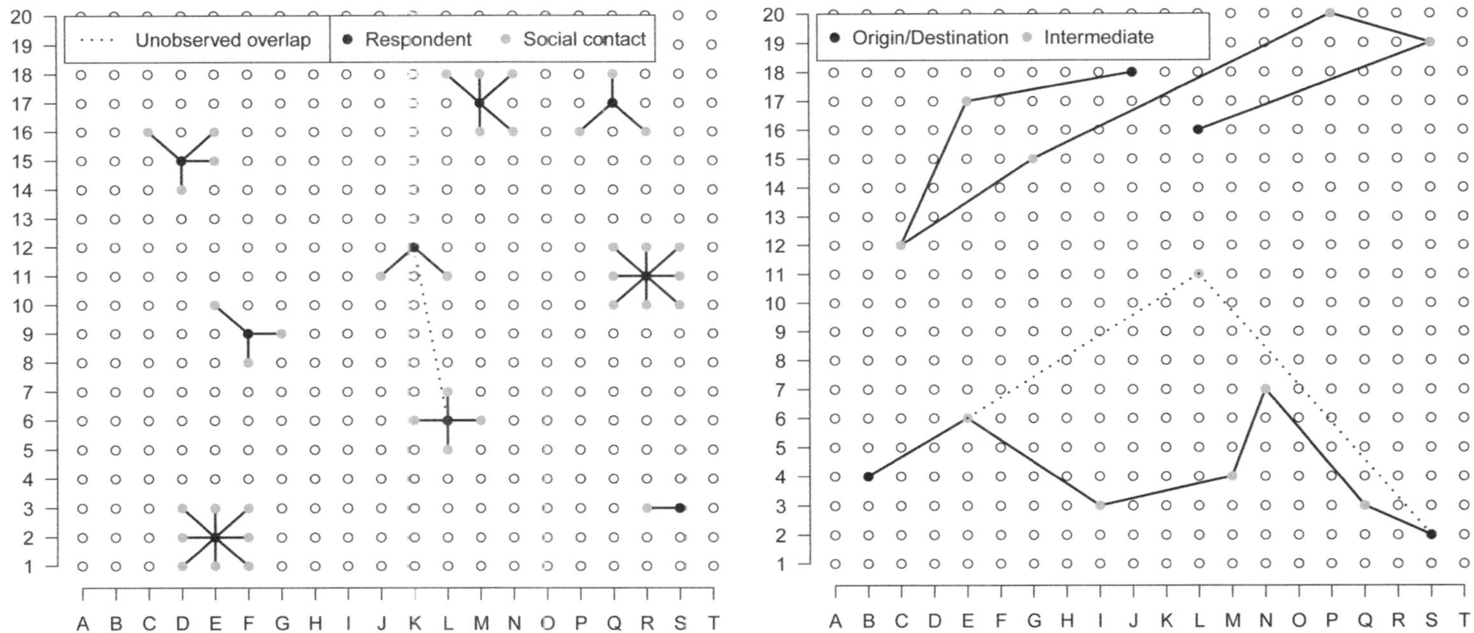

Figure 4.1 Comparison between different sampling strategies a) Isolated personal networks b) Small-world experiment

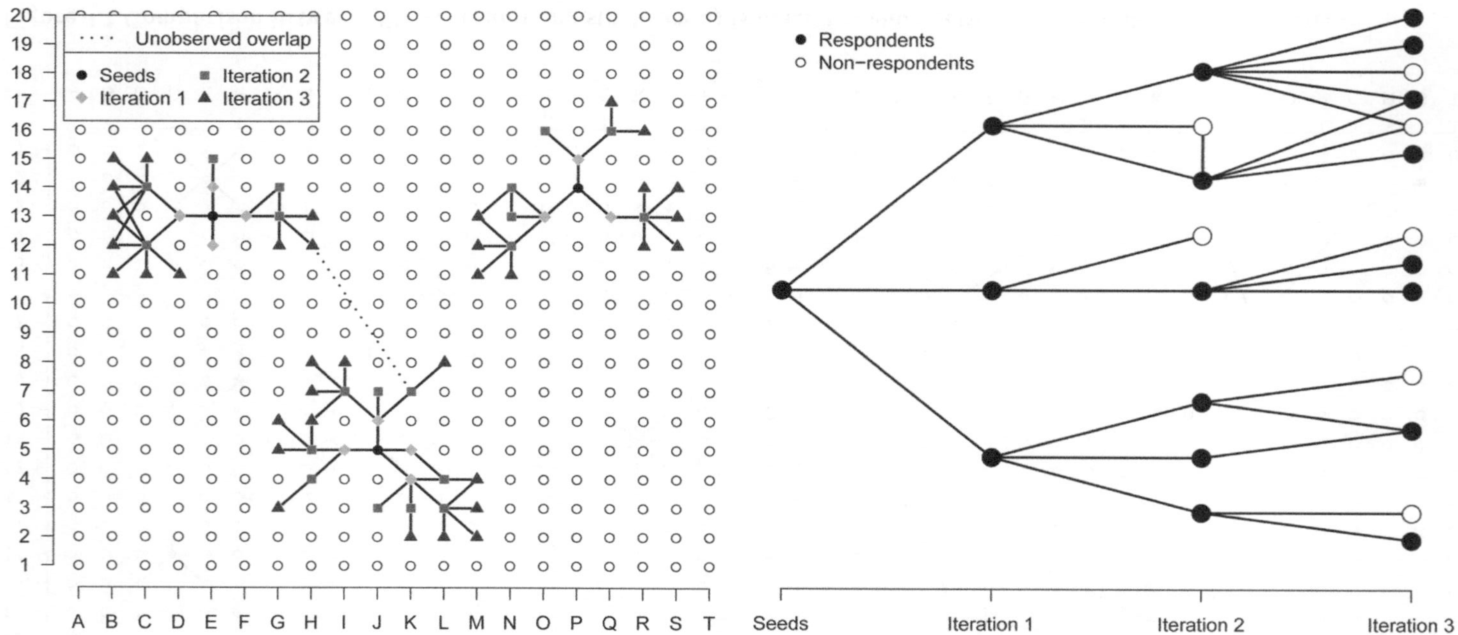

Figure 4.1 (continued) Comparison between different sampling strategies c) Snowball sampling d) Snowball sampling schema

Finally, the snowball method oversamples respondents with larger network structures than average. The more relationships a person has, the higher the probability that this person is asked to participate via one of these communication channels. A team at TU Berlin developed an estimator to correct this '*degree-bias*'. The method weights reported network sizes with their estimated occurrence frequency on each iteration level (for details and references to earlier work on this problem see Illenberger and Flötteröd 2012).

The snowball sampling approach was chosen to satisfy the study aims. To decrease bias, it employed several mechanisms. Chains were started with 40 seeds. Surveying properties of a population-wide network by starting snowball chains with few initial respondents, which are referred to as '*ego-seeds*', and continuing for a medium number of iterations was suggested by a simulation study at TU Berlin (Illenberger and Flötteröd 2012). Continuing with seeds' social contacts, the study collected information on nearly 800 respondents and their approximately 15,600 social contacts. After the ego-seeds' level four additional snowball iterations were undertaken. Project relations between respondents and their social contacts were limited to leisure contacts to best suit the project focus. Researchers imposed no further limitations. Social contacts could live anywhere, did not have to have specific characteristics or use specific communication channels. To our knowledge, a snowball sampling driven study of this size, with so few restrictions, has never been attempted in the field.

Collecting Data: Survey Instrument

The design of a survey instrument is, among other factors, influenced by the investigation's topic, requirements resulting from survey methodology, and attempts to enable easy participation and to keep response burden low. '*Contact diaries*' or '*name generators*' usually collect personal network data. In both methods, respondents are also referred to as '*egos*' and their social contacts as '*alters*'. A name generator is a question that employs stimuli to focus respondents on the part of their social network of interest to a study, rather than collecting information on the entire personal network structure, including all contacts (Marsden 1990; Campbell and Lee 1991; Wolf 2004). While contact diaries can use such a limitation as well, they always imply an additional time boundary limiting the requested period (Fu 2007; for other methods and technologies to survey personal activities see e.g. Bekhor et al. 2013; Zhang and Timmermanns 2012; Makoto et al. 2012). When surveying information on leisure contacts without further limitation, e.g. in time, institutional setting, or communication modes, the name generator approach was preferable to a diary (for a more detailed discussion see Kowald and Axhausen 2014).

The survey instrument had four sections designed to collect detailed information on egos' and alters' personal characteristics and topologies of personal leisure networks. The first section asked for respondents' characteristics, their socio-demographics, and mobility biography (residential and working places throughout their lives).

The second section included two name generators. To survey the specific network as completely as possible, several name generators should be used, providing multiple stimuli. Because name generators ask egos to self-report their network structures, the number of social contacts can be inaccurately reported, caused by forgetfulness, omission, concealment, or invention of alters (Bell et al. 2007). An unsupported collection of network members is thus inadvisable (Pool and Kochen 1978). Furthermore, the number of reported contacts is directly influenced by the number of name generators and space provided to write them down. Instruments including only one name generator (for popular studies, see Wellman 1979; Marsden 1987) are usually biased, because the reported network structures are rather small. However, multiple name generators (the instrument used by Fischer (1982) is a popular one) can also result in bias (for a comparison of name generators and resulting network size see Chapter 3). Possible effects include: fatigue, where people report smaller numbers of contacts the later a name generator question appears, and satisficing where people stop mentioning names when they feel they have reported enough social contacts. Furthermore, multiple questions can result in bias from cognitive priming, when one stimulus overlies a later one, or the more general effect of question-scope redefinition, where former questions influence later ones (Pustejovsky and Spillane 2009). A compromise is a multiple name generator, which uses a moderate number of questions, with each question providing specific stimuli. Following this suggestion, the first name generator of the snowball study asks explicitly for leisure contacts by providing several behavioural stimuli linked to leisure interactions. These stimuli distinguish clearly whether a relationship fits one of the criteria or not. The second name generator uses an affective approach by asking respondents with whom they discuss important problems. This question seeks emotionally important contacts, using a contextual situation with differing possible interpretations instead of interaction stimuli; it could be somewhat imprecise. However, both kinds of social contacts can trigger leisure travel; thus, both name generators were employed. Furthermore, they proved productive in the study of Frei and Axhausen (see Chapter 3). However, question order was changed in comparison to this earlier study to highlight the interest in leisure contacts.

In all, space to report up to 40 names was provided. The first name generator was placed above this space, the second one below. This design was chosen after a pre-test with a different design, where the first name generator provided space for 29 names and the second for 11 additional names. This first design resulted in fatigue and satisficing effects, with a clearly visible artefact in the distribution of reported social contacts. Because many respondents reported 29 names in the first question and none in the latter, the design had to be changed to avoid bias. The re-design (Figure 4.2) solved this problem; the number of reported social contacts increased, on average, although it also had one negative effect. As respondents were asked to report their acquaintances for both questions in the same space, the ability to distinguish between the two kinds of social contacts was lost. Respondents who wanted to report more than 40 contacts were asked to write them down on an extra sheet of paper, so that, in principle, the number of names was unlimited. Still, we assume that at least some respondents refused to make this additional effort due to fatigue. Furthermore, respondents were not asked to report contacts that fit in both generator categories. This helped to decrease response burden but also sacrificed the possibility of multi-relational analyses.

Please list the people with whom you make plans to spend free time. (Examples: errands, sports, club or organized activities, cultural events, cooking together or going out to eat, taking holidays or excursions together.)

No.	Name	No.	Name	No.	Name
Bsp	*Bsp. Doris Musterfrau*	13		27	
		14		28	
1		15		29	
2		16		30	
3		17		31	
4		18		32	
5		19		33	
6		20		34	
7		21		35	
8		22		36	
9		23		37	
10		24		38	
11		25		39	
12		26		40	

If there are other people with whom you discuss important problems, please list them here.

Figure 4.2 Name generators

The third survey instrument section was a 'name interpreter', asking egos to report some basic socio-demographics of each alter mentioned in the name generator and some information on the ego–alter relation, e.g. duration and circumstances of first meeting. Due to the ascending nature of snowball sampling, the instrument also collected information on social contacts' addresses. Participants not wishing their network members to be contacted for further recruitment were asked to write this down, or to use anonymous nicknames for their alters. The name interpreter also included a check box, which respondents had to mark if they wanted to stay anonymous in further recruitment. This meant they did not wish their alters to know who mentioned their names.

The fourth and last section of the instrument was a 'sociogram', asking respondents to mention those social contacts from the name generator who make plans to spend free time together. Sociograms show a wide range of diversity, from socio-matrices, which ask egos to mark any pair of alters that have a certain kind of relation, to visualizations, measuring different kinds of relations and their emotional closeness. They can be used as a form of a name generator (Hogan at al. 2007) or in addition to a name generator, which is the case here. While socio-metrical tables are often considered boring and result in fatigue effects, graphical elements imply theoretically driven categories, which can influence respondents' answers (Mamali 2006). The snowball study sociogram was designed to collect information on cliques within the personal network topologies – specifically alters making plans to spend free time together. This snowball study dealt with people all over the world; an interviewer could not always be present, meaning that the design had to allow easy participation. The current design addressed these requirements by asking respondents to mention contexts in which their social contacts meet, e.g. birthday party or hiking group, and to list all alters joining one of these meetings regularly. To reduce response burden, names could be reported with the help of a predetermined shortcut from the name generator. However, respondents were not allowed to mention any social contacts beyond those reported in the name generator. While this kind of sociogram allows easy participation, it automatically results in fully connected clique structures. It implicitly assumes that all alters joining one leisure group known to the ego are also fully known to each other. The assumption of fully connected cliques is thus a potential source of bias, particularly significant in a large clique. Nevertheless, the present design is a good starting point for leisure network topologies: as almost all cliques are rather small, presumably an accurate one as well. Respondents were allowed to mention up to 20 cliques; thus, each alter could be a member in more than one clique.

All questions in the instrument were formulated in a clear, easily understandable and unambiguous way. Furthermore, examples were given wherever questions could have caused problems. This supported respondents in comprehending, judging, and finally answering the questions (Groves et al. 2009; for detailed information on developing and pre-testing the survey instrument see Kowald et al. 2009). The survey instrument was issued exclusively as a paper version in German and English. Although an electronic version was available, it was not used in the snowball process because of doubts about its operation with various browsers. In addition, lower response rates (Sen and Bricka 2009), and higher rates of drop out behavior and item non-response are reported for online instruments (Heerwegh and Loosveldt 2008; Couper and Miller 2008; for problems with online surveys on personal networks see Matzat and Snijders, 2010). Given these issues, we decided to keep high technology in the lab and use low technology in the field (Hogan 2007).

Reducing Bias: Survey Protocol

Conducting a snowball sample without geographical limitation means that respondents and their social contacts could live anywhere. Chains were started in Canton Zurich, Switzerland and previous studies showed that most leisure contacts are local or regional (see Chapter 3). So, it was expected that egos and alters would be clustered in Switzerland, which indeed was observed (Kowald and Axhausen 2014).

Employing snowball sampling and name generators introduced several biases. In order to decrease bias, the study adopted several measures: one related to the fact that snowball sampling uses only a few seeds to start its chains. According to the statistical law of large numbers, this can result in bias, as there is no guarantee that these few seeds represent target population characteristics, even when recruiting from a random sample. To improve the chances of covering at least some personal characteristics distributions properly, the survey used a stratified random sample of the Canton Zurich population to recruit its seeds. This part of Switzerland was chosen because it exemplifies Swiss population heterogeneity and includes urban and rural areas. The sample was stratified in terms of sex and age and matched the share of people living in urban or rural environments. While over- or under-sampling bias is already a problem when employing 'usual' sampling techniques, it is even worse for snowball sampling. Due to a frequently observed similarity between respondents and their social contacts, this bias might not only be passed on from iteration to iteration, but even increase during this process, a phenomenon we called '*homophily-bias*' earlier. To have a chance of balancing homophily-bias, the sample employed a low-tech approach: snowball chains were started in two consecutive sub-samples. Each sub-sample used 20 seeds. After data collection on the final iteration level was completed for sub-sample 1, data were analysed to see if the sample had been captured in a homogenous cluster. Recognizing that this had indeed occurred with egos' ages triggered the decision to start the second sub-sample by selecting younger seeds. Ultimately, this helped to balance bias and cover the Swiss age distribution more accurately, as a comparison to data from the Swiss Microcensus on Mobility and Transport (MCMT), a weighted sample to represent the Swiss population, shows (see Table 4.1).

Table 4.1 Using consecutive sub-samples to balance sampling bias

Age categories	Egos Sub-sample 1		Egos Sub-sample 2		All alters		MCMT	
	% obs.	mean age	% obs.	mean age	% obs.	mean age	% obs.	mean age
21–40	17.9	33.4	17.6	32.3	22.0	32.5	28.0	31
41–60	35.3	49.8	63.1	49.6	49.8	49.6	31.4	50
61–80	42.4	68.9	17.8	66.8	22.8	68.1	18.4	69
81 +	3.5	82.3	1.1	83.0	2.4	84.4	4.3	85

Source: For the Swiss Microcensus on Mobility and Transport (MCMT) see ARE/BFS (2012).

Seeds were important for another reason as well. Asking them to fill out the instrument with the help of interviewers was an attempt to improve data quality and create a basis for comparison with data collected by mail questionnaires on later iteration levels. Interviewers were trained to ask the name generators questions repeatedly to make the seeds mention their contacts completely. In addition, they were trained to answer questions about the aims of the project and deal with respondents' concerns on data security and anonymity.

This is related to the second major bias concern: uncertainty whether respondents mention all their leisure contacts and whether these contacts can be recruited as completely as desired. In other words: studies combining snowball sampling and name generators should include measures to give respondents confidence when mentioning their contacts. This helps decrease selection bias and increases response rates to obtain a more complete picture of the underlying global network structure.

Textbooks on quantitative research studies suggest using a '*multi-contact strategy*' to establish and maintain contact between potential respondents and the research team (Dillman 2000). This proves researchers' interest and efforts in collecting data. Furthermore, each contact can be an opportunity to inform respondents about study aims, issues regarding their anonymity, their status as volunteer participants and how to handle the survey instrument. Following up on this idea, the study employed an announcement letter, a recruitment call, an accompanying letter sent with the questionnaire and a prepaid envelope to return the instrument, a remainder letter for those who did not return the survey instrument within three weeks (if needed), and, finally, a thank you letter. In addition, a web page was designed providing detailed information on the survey, the research team, and links to all publications resulting from the project. Besides documenting researchers' efforts in collecting data, both multi contact strategy and web page increased the transparency of the study, created a trustworthy atmosphere and made potential respondents feel confident about participating. All contacts employed mechanisms from goal framing theory, an enhancement of the traditional rational choice approach. Both the multi-contact strategy and web page tried to prime potential respondents in a normative way. Asking for help in developing a new scientific approach emphasized the study's scientific character. Assuming that people are often primed in a hedonic way, this shift increased the probability of potential egos to take on the response burden and fill out the questionnaire (for details on goal framing theory see Chartrand and Bargh 1999; Lindenberg 2006; Lindenberg 2008; Aarts et al. 2004).

Furthermore, we tried to express our appreciation of respondents' efforts in filling out the survey instrument. Following the idea of a '*gratification strategy*', 20 CHF (or the equivalent amount in the currency of potential respondents' residence country) were sent with the survey instrument. This sum seemed appropriate, as the survey instrument included a substantial response burden and most respondents lived in Switzerland, where 20 CHF could fund, e.g. an evening in a cinema. Money was preferred to non-monetary gifts or gift certificates, as this kind of appreciation is understood universally. Whenever a gratification strategy is used, the token should be sent with the survey instrument to boost the response rate. Dillman (2000) argues that the appreciation transforms the situation between potential respondents and research team into a social exchange, functioning as social approval given in a symbolized form, encouraging people to behave in a normative, rather than a hedonistic, way (Frey 1997; Diekmann and Jann 2001). However, sending a token after the fact usually does not produce significant results; it is more of an economic exchange, which would be like paying respondents for their participation. Regarding gratification strategy, it must be mentioned that respondents understood their participation was voluntary and, of course, non-respondents were allowed to keep the token without any pressure or obligation. They did receive a reminder letter asking them again to participate, but the token was not mentioned.

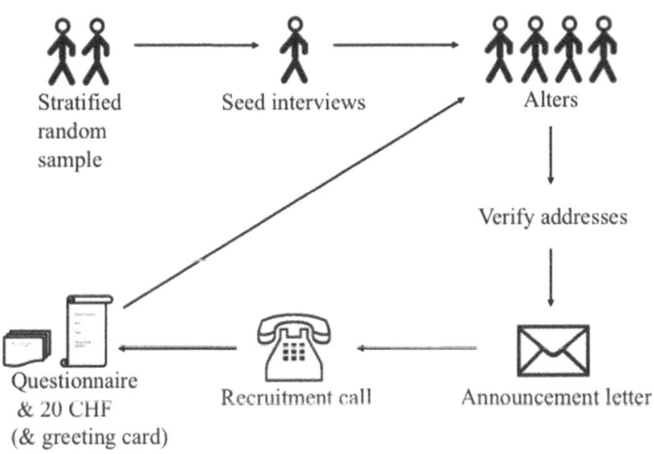

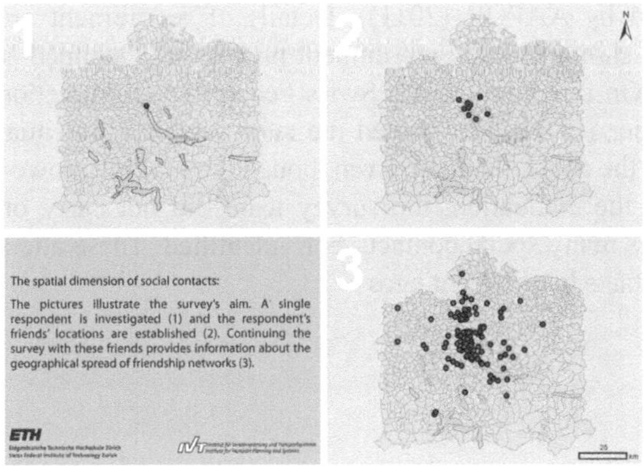

Text on the backside of the card:

I took part in a survey of ETH Zurich regarding my social network. In answering their questions I mentioned your name. I'm sending you this card to alert you that ETH will contact you and also to ask you to take part in the project.

Figure 4.3 Survey protocol and greeting postcard

Finally, the survey included a chain method-related strategy to increase the response rate. Using previous respondents to recruit new participants is a strategy in respondent-driven sampling, where respondents receive a token for filling out the questionnaire and a second one for recruiting their social contacts (Heckathorn 1997). Here, egos were asked to fill out a prepared greeting card in postcard format and either return it to the survey team, or send it to their alters. This card confirmed to potential respondents that a trusted contact had already participated in the study, increasing the chances of participation. In other words, the card was an attempt to initialize goal contagion behaviour, which is the strategy of copying behaviour, often observed between closely related persons. Both effects, achieving a normative frame and goal contagion, are extraordinarily strong when they occur together (Cialdini et al. 1990). When respondents used the cards, they lost their anonymity in the recruitment process, meaning that their alters were able to ascertain who mentioned their names. This was the case for around 10% of all alters. However, if respondents did not use the cards, they remained anonymous in further recruitment.

Field Work

The snowball methodology necessitated checking all names to avoid re-sampling, which could have happened when egos mentioned alters already in the sample. Avoiding re-sampling was also necessary for an accurate picture of the underlying global network, or re-sampled persons would have been captured as two different individuals. Student assistants did this work by comparing names and postal addresses. The combination of multi-contact strategy, incentive, and name comparisons was time consuming and resulted in survey costs of around 110 CHF per respondent (see Kowald and Axhausen 2014).

These measures helped increase response rate. Overall, the survey achieved a response rate of around 20% (calculated as the RR1, valid addresses divided by participants, as suggested by AAPOR (2011)). Details of recruitment are provided in Table 4.2, which shows that the recruitment process was stopped at the beginning of iteration 4. On this iteration the Swiss Federal Data Protection and Information Commissioner (FDPIC) evaluated the survey after a potential respondent complained about the data collection. Even though the FDPIC allowed to continue the survey after the evaluation, the survey team did not carry on however. Table 4.2 also shows many social contacts as re-identified. These alters were mentioned as social contacts by different egos.

Table 4.2 Recruitment process

	Whole Sample		Seeds	Iteration 1	Iteration 2	Iteration 3	Iteration 4
	Abs.	[%]	Abs.	Abs.	Abs.	Abs.	Abs.
Sample size	12,940		285	568	1,828	4,073	6,186
Re-identified	1,376		–	–	247	444	685
Valid addresses	4,138	100.0	247	385	1,186	1,343	977
Participated	793	19.2	40	119	276	291	67
With recruitment	711	17.2	40	105	249	257	60
Without recruitment	82	2.0	–	14	27	34	7

Considering the unusual survey methodology, high amount of response burden and level of confidentiality implied in the questions, the achieved response rate is satisfying. A rating system from commercial survey research assesses response burden by weighting each question according to its structure, predicts an instrument's response rate and helps qualify the multi-contact strategy (Axhausen and Weis 2010). Figure 4.4 shows the cooperation rate for two sub-samples from the survey; both include persons who received the questionnaire. The sub-sample 'with recruitment' received the whole multi-contact strategy, while those 'without recruitment' did not receive every step of the multi-contact strategy because no valid phone number was available for them. Instead, this second group received an announcement letter and, one week later, the questionnaire with the incentive and (if provided) the greeting card. The effect of the full multi contact strategy is clear. Persons 'with recruitment' show a cooperation rate of over 70%: those 'without recruitment' a cooperation rate of approximately 25%. The cooperation rate was calculated as the COOP4, number of persons agreeing to participate divided by participants, as suggested by AAPOR (2011).

Both multi-contact strategy and greeting card can be used as explanatory variables in a 'decision to participate or to refuse' logistic regression model. The estimate shows the multi-contact strategy to be highly effective. Those respondents who received every contact are twice as likely to participate as those who received only partial contacts. The greeting card also has a substantial positive influence on participation.

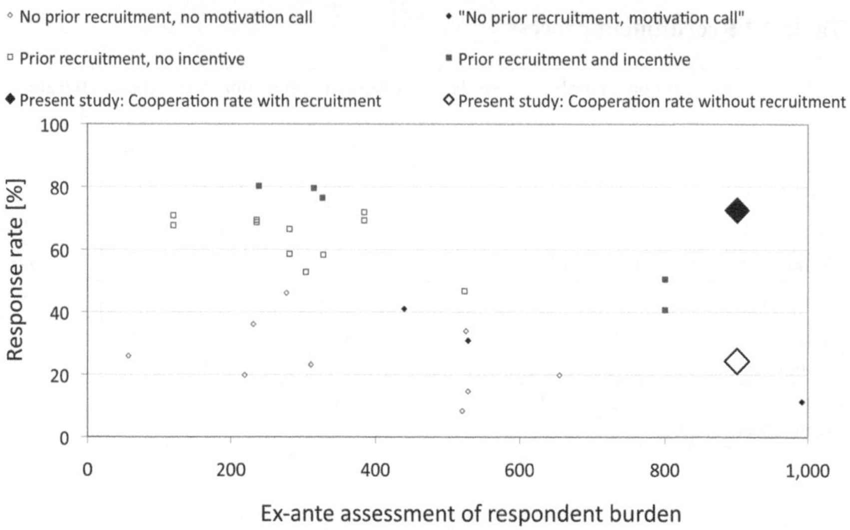

Figure 4.4 Effects of the multi-contact strategy

Table 4.3 The influence of multi-contact strategy and greeting card on response rate

Coefficients	Estimate	Std. error	p-value
Intercept	−1.306	0.127	<0.001
Dummy full multi-contact strategy	2.038	0.144	<0.001
Dummy postcard	0.392	0.157	0.013
Null deviance	1,845.6		
Residual deviance	1,588.3		df= 1,344

Modelling the incentive strategy effect is more difficult, as it was sent to all potential respondents. However, analysing data from recruitment shows surprising results; a significant share of both respondents and non-respondents sent back the incentive. Employing a pivot table and a chi^2–independence test shows a significantly different distribution than expected (chi^2=139.1; p=<0.001). Around 9% of all respondents and 34% of all non-respondents returned the incentive (see Table 4.4).

Employing a logistic regression model on the decision whether to keep or return the incentive also indicates that this behaviour is influenced by the multi-contact strategy. Having received all contacts increases the probability of returning the incentive. A negative effect results from response status (respondents or non-respondents) showing participants being less likely to return the incentive. The same is true when having received a greeting card (as the card increases the probability for participation). Socio-demographics are largely irrelevant, with the exception of age (we also checked marital status, level of education and sex). As respondents age, they are more likely to send back the incentive. The model is summarized in Table 4.5.

Table 4.4 Incentive and response behaviour

	Respondents	Non-respondents	Marginal frequencies
Incentive kept	728 (52.8%)	385 (27.9%)	1,113 (80.7%)
Incentive returned	68 (4.9%)	199 (14.4%)	267 (19.3%)
Marginal frequencies	796 (57.7%)	584 (42.3%)	n = 1,380 (100%)

Table 4.5 Logistic regression model on sending back the incentive

Coefficients	Estimate	Std. error	p-value
Intercept	−2.017	0.318	<0.001
Respondents	−1.896	0.171	<0.001
Postcard sent	−0.289	0.203	0.154
Full multi-contact strategy performed	0.756	0.176	<0.001
Age [years]	0.018	0.006	0.002
Null deviance	1,304.2		
Residual deviance	1,155.5		n = 1,335

Using snowball sampling in a study of this size on personal leisure networks raises issues of data protection and ethical considerations. In the field, the survey was confronted with three major concerns. All of them were checked by the Swiss Federal Data Protection and Information Commissioner (FDPIC). Naturally, personal data collection must always address issues of data security and anonymity. Furthermore, respondents must be made aware of their status as volunteer participants (even more crucial when using an incentive strategy), and the project goals.

First, like most studies on personal networks, the survey instrument included a name interpreter. This entails asking respondents about characteristics of third parties – their alters. Respondents usually have no idea that they (and some of their personal details) have been reported unless the ego tells them (Kadushin 2005; Klovdahl 2005). This problem is very acute in a snowball sample. The necessity of collecting data is accentuated here because alters' names and postal addresses are necessary to continue recruitment. However, the problem of collecting information on third parties is less critical in a snowball sample because all alters (at least those for whom valid addresses were available and for whom the ego allowed a contact) were asked to participate in the survey. So, almost all alters were informed about the survey study collecting data on them. Should an alter want information about which personal data were collected, this information had to be provided. In case the person wanted the data deleted, this had to be done as soon as possible. Both cases occurred in the present sample, but with 3 out of around 15,500 observations, only as very exceptional events.

Issues of data protection and anonymity are also relevant to a second topic. Usually, data are collected to allow calculation of anonymous statistics. In Switzerland, it is forbidden by law to collect personal data and use them in a personalized way. However, for this project, it was necessary to handle data both anonymously and personally. All alters had to be checked for re-identification and overlap between network components to avoid re-sampling. By employing two databases, one for recruitment and one for analysis, permanent distinction between these data was maintained. Only the recruitment database included names and addresses and they were deleted after the study. This separation made it feasible to use snowball sampling in Switzerland.

Regarding egos' anonymity, the Swiss Federal Data Protection and Information Commissioner (FDPIC) disagreed with the approach used in our study. If egos wished to remain anonymous, or did not sign greeting cards, their alters were not told who provided their name, even if they requested this information. Swiss law on data protection allows such exceptions to the general obligation of informing people what kind of data were collected on them and how this information was gathered, if the privacy of a third person is endangered. The FDPIC decided, however, that egos' wish to stay unmentioned is not sufficient reason for such exceptions. As this assessment was published after the study had been in the field, this project was not affected, but the topic must be addressed in future survey studies.

Representativeness and Data Quality

One particular challenge in a snowball sample is defining the '*inference population*' for which results from the study should be valid. The snowball study aimed to survey characteristics of a population-wide leisure network. As snowball chains were started in Switzerland and as most leisure contacts were expected to be of local character, the survey team assumed that it would collect data on the Swiss leisure network. However, egos were allowed to mention alters located elsewhere and these alters were used to continue snowball chains. Therefore, it was unknown which geographical patterns would be observed and where snowball chains would spread to (see Kowald and Axhausen 2014).

Exemplifying Swiss population heterogeneity and including urban and rural areas was the aim of recruiting seeds from a stratified random sample of the Canton Zurich population. The initial sample included people drawn randomly from the telephone directory. So, the '*frame population*', that is all the people who had the chance of participating in the study, was on this level limited to those with a registered address and a record in the telephone directory of Canton Zurich. However, as snowball sampling means that all non-seed respondents are introduced by former respondents, other groups of people were excluded from the frame population (on the later iteration levels) too. First of all, there was no chance for people without leisure or emotionally important contacts to become part of the sample (with the exception of the seeds). Due to missing contacts, such '*isolates*' are not mentioned by others and therefore excluded from the frame population. Secondly, people belonging to hard-to-reach sub-populations had a lower chance of becoming part of the sample. This includes people that fear discrimination or contact with governmental authorities, like e.g. drug addicts or illegal immigrants.

The '*survey population*' includes all people who become part of the sample. As mentioned above, snowball chains were started with 40 seeds and finally resulted in a sample in which nearly 800 egos mentioned around 15,600 alters. Both, egos and alters are clustered in the German speaking part of Switzerland and particularly in Canton Zurich. However, there are both egos and alters in other parts of Switzerland as well and only very few people in other countries (less than 1% of all egos; for the spatial distribution of egos' home locations see Figure 4.5).

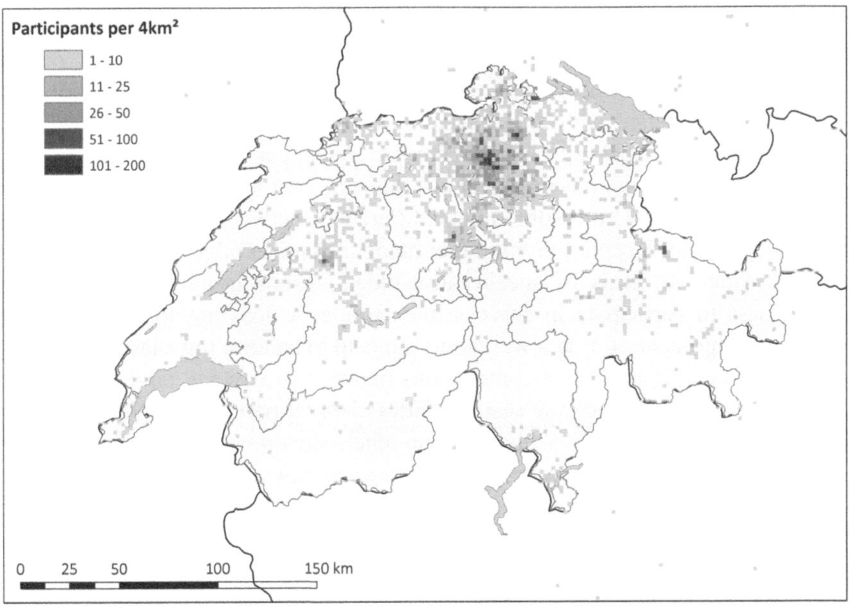

Figure 4.5 Egos' home locations (per 4 km²)

Due to the exploratory character of the study and the unknown spatial distribution of the snowball sample the survey team abstained from defining a target population before the survey was in the field. However, we expected to survey properties of the Swiss-wide or at least the leisure network of the German speaking part of Switzerland. This expectation was fulfilled in terms of geography and the survey team decided to define the Swiss population as the '*target population*'. The target population includes those people who had the chance of becoming part of the frame population.

A comparison between the survey population and the population of Switzerland shows a good fit. Bearing in mind that snowball sampling does not match the criterions of a random sample and that it includes several sources of bias, this is rather surprising. However, there are some cases of over- and under-representation.

In terms of sex an 11% over-representation of females can be seen for egos and a 7% over-representation for alters. In terms of civil status the egos show an under-representation of singles of around 20% and an over-representation of married persons of 18%. Other differences are rather small, around 1%, on the other factor levels. Focussing on alters shows smaller differences: Singles are under-represented by around 11% and married persons are over-represented by 11%. Distribution in civil status is correlated with the age distribution. In the snowball sample youngsters (0–20 years) and young adults (21–40 years) are under-represented, whilst middle aged persons (41–60 years) are over-represented by around 24% for egos and 19% for alters. For egos this difference results in part from the sampling strategy as people under 18 years were not asked to participate. Due to homophily bias, this mismatch is passed on to the alters: Where people can choose their relationships they usually show a preference for others with similar characteristics. In this sense the difference between the alters and the Swiss population is due to the sampling strategy too. A comparison between the survey- and the Swiss population is provided in Table 4.6.

In summary, there are some major differences between the survey- and the target population. Keeping in mind that random samples often show mismatches as well, often in terms of age because older persons usually show a higher willingness to participate in surveys, the sample fit does not look serious and statistical approaches for data weighting can help to balance this bias. In addition, one has to keep in mind that comparisons between survey and target population are always limited to certain characteristics. This is particularly challenging in case of the snowball study as there are no other network studies, which could be used for comparisons of network issues. However, wherever possible we applied such comparisons and they showed the snowball data as being plausible (Kowald 2013; Kowald et al. 2013). Therefore, and due to the fact that there are no other studies on the Swiss population-wide leisure network, the survey team decided that the statistical results are useable for the Swiss population by keeping in mind the explorative character of the study.

Table 4.6 Comparisons between egos' and alters' characteristics and the MCMT

Character	All egos (n = 793)		All alters (n = 15'593)		MCMT Switzerland (n = 62'868)	
Sex [%]						
Male	38.1		42.0		49.0	
Female	61.9		58.0		51.0	
Civil status [%]						
Single	13.4		22.2		32.6	
Married	70.9		64.1		52.9	
Divorced	9.1		8.1		8.4	
Widowed	5.3		4.5		6.1	
Married, living separately	1.3		1.1		0.0	
Age	**Share [%]**	**Mean age**	**Share [%]**	**Mean age**	**Share [%]**	**Mean age**
0–20	0.5	19.0	3.0	14.7	17.9	13.5
21–40	17.7	32.7	22.1	32.5	28.0	31.2
41–60	55.0	49.7	49.8	49.6	31.4	49.9
61–80	25.0	67.8	22.7	68.1	18.4	69.3
81+	1.8	82.6	2.4	84.4	4.3	84.9

Source: For the Swiss Microcensus on Mobility and Transport (MCMT) see ARE/BFS (2012).

The term '*item-non-response*' referes to missing values in particular items of the survey instrument. Item-non-response includes questions which respondents refuse to answer because they either do not understand the question, do not know the answer, or do not want to provide information, e.g. because of privacy concerns or too high a response burden. In case of snowball sampling item-non-response also describes egos who reported several alters but do not report particular characteristics for one or several of these social contacts.

In a name generator guided study on social networks it is important to distinguish between different network sizes when focussing on item-non-response behaviour as each single alter mentioned in the name generator leads to follow-up questions with related response burden in the name interpreter. So, respondents with many alters have a higher response burden than those with few reported social contacts.

The shares of missing values per question cover a wide rage (between 0 and 30%). They differ between questions and parts of the questionnaire. Unfortunately, the survey instrument does not allow judgement on whether a missing value corresponds to item non-response behavior, is the answer of a person who does not match any attribute of the question and thus cannot provide an answer, or is unsure how to answer it. Although literature on survey instruments recommends a design allowing this distinction (see e.g. Noelle 1967; Dillman 2000; Diekmann 2001) we did not use such additional answer opportunities to keep the response burden low and the survey instrument short. Questions on egos' and alters' educational level (questions 12 and 29) where an increased share of missing values can be observed (around 10%) in particular could be affected by this shortfall. As response burden is low for these question there must be another explanation. It might be possibile that respondents from countries others than Switzerland and Germany or immigrants in Switzerland with other educational systems were not sure how to answer the question or which factor level summarizes their own educational experience. Including answer categories that allow a distinction between refused answers and respondents that do not know an answer would have addressed this issue.

Shares of missing values for different network sizes are shown in Figure 4.6. In general, non-response becomes more frequent when questions are asking for hard-to-remember information like egos' mobility biography (question 20 and 21) or year and context of the first meeting with an alter (questions 31 and 32). Second, an increased share of missing values is observed in questions asking for confidential information like names and particularly addresses of social contacts (questions 23 and 24). Finally, an above average share of missing values occurs whenever a question design can be considered unusual like the questions on communication modes and -frequencies (questions 33 to 36). These later items asked for the frequency of communication within a certain time frame. To answer it respondents had to choose the time unit that best fits the frequency they wanted to report from a predetermined list: per day, per week, per month, or per year. In combination with the communication frequency, e.g. two times a day, 20 times a month, or 50 times a year, this allowed the calculation of annual communication frequencies. The idea was to allow an easy and intuitive report of communication frequencies without making calculations necessary on the respondent's side. However, the increased share of missing values is a hint that at least some respondents might have been confused by this design. In addition, some participants may have lost interest when answering these similar and consecutive questions on communication behaviour. Aside from hard-to-remember or confidential information and unusual question designs both question position and amount of response burden clearly influence the share of missing values. Using a Spearman's test to check the correlation between response burden and question position and related shares of missing answers shows the relation being positive, 0.459, and significant ($p = 0.003$) for response burden and highly positive, 0.801, and significant ($p = 0.000$) for question position.

Figure 4.6 shows a surprising result for different network sizes (egos with 1 to 10 alters are plotted in the top graph, egos with 11 to 20 alters in the 2nd graph, respondents with 21 to 30 alters in the 3rd graph, and, finally, egos mentioning 31 to 40 alters in the bottom graph. Each plot includes a network size specific moving average, the black line, and a moving average over all network sizes, the bright line. Both smoothing functions were calculated as the average of missing value shares over five consecutive variables. The vertical dotted lines represent bondaries between different parts of the questionnaire, whereby 'NG' means name generator and 'SG' sociogram). Intuitively it could be assumed that non-response behavior is positively correlated with an increasing network size due to fatigue effects. Indeed, this tendency is present in the data as the overall moving average increases with each additional question. But there is also another tendency, which seems to be related to network size: Respondents' motivation to answer questions.

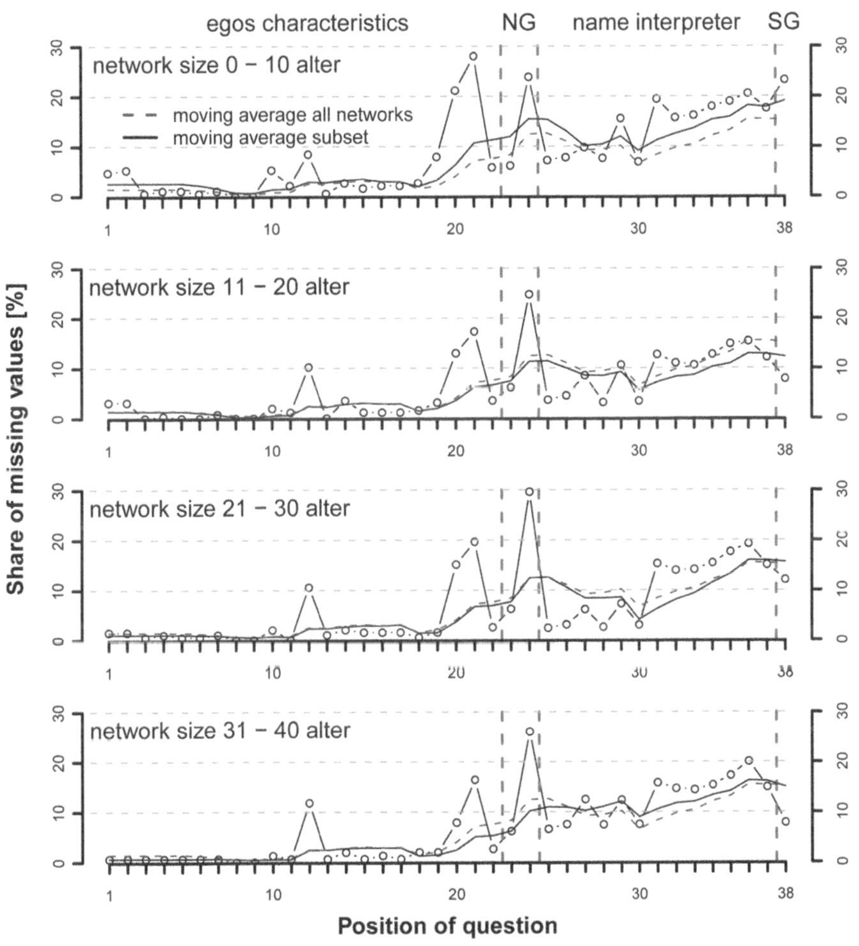

Figure 4.6 Share of missing values (NG = name generator; SG = sociogram)

Table 4.7 Egocentric networks attributes; only egos with completed sociograms (n= 639)

	Mean	Std. deviation	Minimum	Maximum	25-percentile	Median	75-percentile
Number of nodes	21.92	10.10	2.00	40.00	14.00	20.00	29.00
Number of ties	44.75	57.49	1.00	378.00	10.00	24.00	54.00
Number of isolates	7.41	6.60	0.00	34.00	2.00	6.00	11.00
Number of cliques	4.09	2.51	1.00	20.00	2.00	4.00	5.00
Members per clique (mean)	4.44	2.35	2.00	28.00	3.00	4.00	5.33
Number of components	2.56	1.44	1.00	9.00	1.00	2.00	3.00
Density	0.19	0.18	0.00	1.00	0.07	0.14	0.25
Density without isolates	0.43	0.26	0.09	1.00	0.24	0.36	0.50
Degree centralization	0.22	0.18	0.00	1.00	0.10	0.17	0.22
Betweenness centralization	0.11	0.18	0.00	1.00	0.00	0.01	0.16

Those 189 egos (23.8%) mentioning small networks with between 0 and 10 alters (top graph) are weakly motivated. They already start with an increased share of missing values when compared to the overall average. This value further increases more or less with every additional question. Especially when facing challanging questions with a high response burden these respondents often reacted with item-non-response behavior. In nearly all questions the within-network-size line is higher and smoother than the average over all networks.

Less variance can be observed for the 254 (32.0%) respondents with network sizes between 11 and 20 alters (2nd graph). It is worth noticing that this category is the densest in terms of respondents and contains the average network size not only of the present project but also of survey studies asking for similar kinds of information (see e.g. Carrasco 2006; also see Chapter 3). Overall respondents within this category start with low shares of missing values. There are also only a few missing answers in the name interpreter. These participants can therefore be labeled as highly motivated to answer all questions completely. The class size specific moving average is nearly always lower than the moving average over all network structures.

The third category of network sizes contains 198 (25.0%) egos (3rd graph). Starting with a low share of missing values the non-response behavior of this group stays low for the egos' characteristics. Unlike respondents in the second class of network size, however, this behaviour does not last for the whole questionnaire. Starting at question 31 (the year the ego and alter first met) the share of missing values increase rapidly and stay high for the following questions. It seems that this group loses interest in filling out the questionnaire at some point in the name interpreter. Alters and questions lying beyond this point have a high probability of remaining unanswered. Because missing values increase rapidly in the name interpreter, the behavior of this group can be described as motivated but fatiguing. This tendency can also be seen when comparing the class specific moving average with its counterpart over all networks.

A similar but somehow stronger tendency is observed for the fourth group of 152 (19.2%) egos with large network sizes (4th graph). This group also starts with low shares of missing values and, similar to group three, shows a benchmark in the name interpreter beyond which questions have an increased chance of remaining unanswered. Comparing the within-group smoother to the sliding average over all groups confirms this finding. However, considering the high level of response burden resulting from the high number of reported social contacts, the overall amount of missing values still looks reasonably good, when compared to the other groups. So, group members show a strong willingness to answer questions. In this sense the within-group behavior can be labeled as idealistic.

In summary, a question's position and its resonse burden significantly influence participants' response behavior. Furthermore, questions which are difficult to answer, whether because they include difficult recall tasks, ask for confidential information, or show an unusual design, result in increased shares of missing values. So, they should be avoided or, at least, reduced to an essential minimum.

Questions dealing with similar topics and being of similar designs should not be asked consecutively as respondents get bored quickly. In addition, whenever answer categories for a particular question might result in problems for (a subset of) respondents it is necessary to include additional answer categories allowing a distinction between those refusing to answer, those that do not match any category, and those who do not know the answer. Finally, respondents' motivation is of high importance. In comparison to the other findings this issue is hard to address as it does not depend on the reserchers' efforts alone.

Empirical Results: Personal Networks

The term '*personal networks*' describes all data reported by egos on their egocentric network structures. So, the following analyses focus on a subset of the snowball data. Only isolated components of personal network structures are considered. Any overlaps captured between these structures are ignored here.

In general, the snowball data allows a repetition of all the analysis from Chapter 3. However, instead of reproducing the descriptive statistics and model results on distances between egos and alters and their mode specific communication frequencies from Chapter 3, the analysis here focuses on additional research topics, which have been treated only marginally in transport planning so far: homophily and tie strength. However, first it is necessary to give an idea of personal network structures (readers who are interested in comparisons between surveys on personal networks in general and the distance distribution between social contacts in particular can find such a study in Chapter 5).

Figure 4.7 shows an empirical example of a network to illustrate how personal networks from the snowball data are analysed. The example includes an ego who reported 14 social contacts (Figure 4.7a). In graph theory each person (ego and alters) is called a '*node*' and each relationship is called a '*link*'. Terms like associates, leisure contacts, and persons are used as substitutes for 'nodes' here. The sociogram data show that most alters in the example are organized in one out of four leisure groups. One alter even participates in two groups and thus creates an overlap between them. Each of the groups is fully connected, which is due to the survey instrument: All nodes within a group have ties with each other. The graph also shows two alters that do not participate in any leisure group (Figure 4.7b). Usually, egos and ego – alter relations are excluded from the analysis because, following the logic of the egocentric network approach, the egos would always be the most central persons in their network structures and all statistical figures would be biased by them. The resulting Figure (4.7c) introduces some additional terms from graph theory: '*Cliques*' are referred to as maximum connected sub-graphs, the groups reported. In a clique all members are adjacent and directly connected. In contrast, a '*component*' demands only a path between all nodes in the component and no direct connection between all nodes belonging to the component. In other words, component members are connected

a) Network structure from name generator

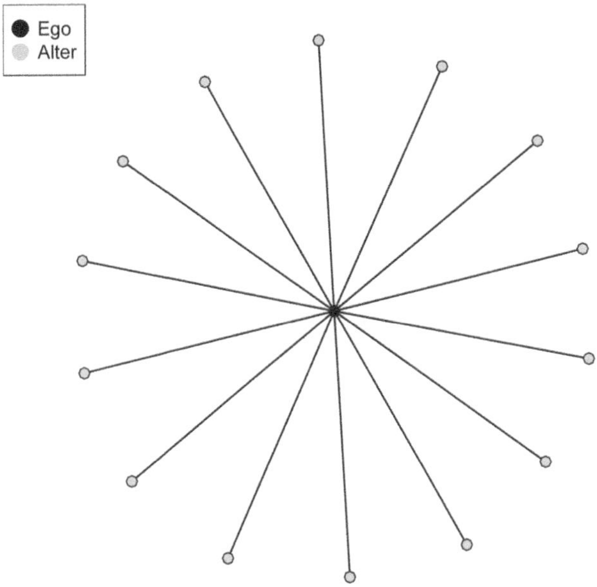

b) Network with sociogram information

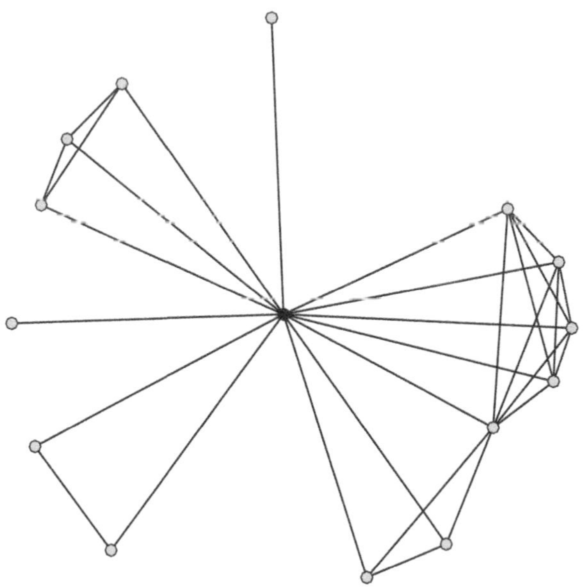

c) Network without ego and ego-alter ties

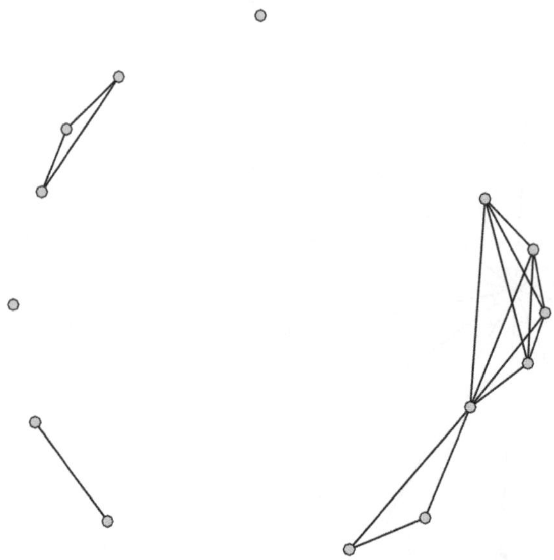

d) Network without ego, ego-alter ties and isolates

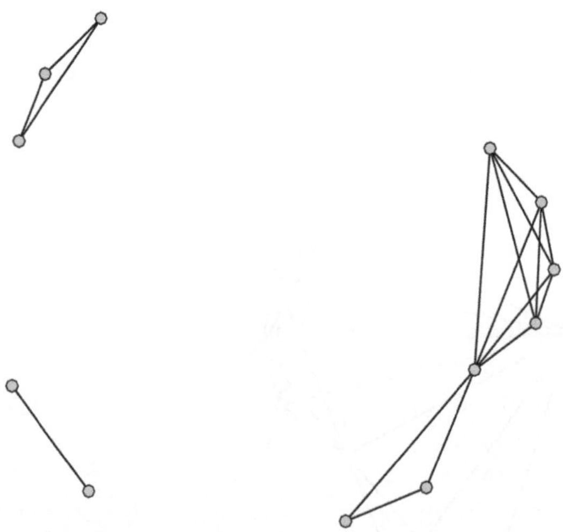

Figure 4.7 Empirical showcase for a personal networks analysis

by direct or indirect links. According to that definition, the example shows four cliques and three components where the individual that joins two cliques, and thus creates an overlap, connects these two cliques into one component. In addition, the '*isolates*', those completely unconnected nodes, are each a component on their own. To calculate the number of cliques and components including more than one person the isolates are excluded to avoid bias (Figure 4.7d; for further information on graph theory see Jansen, 2003; Wasserman and Faust, 2007; Scott, 2007). In the analysis all links are considered as being undirected and multi-relational analysis is relinquished. So, if an ego mentions an alter as a leisure contact, the ego is also a leisure contact of the alter. Furthermore, if information from the sociogram indicates that two alters meet in several leisure activities (e.g., tennis and at the ego's birthday parties) only one relationship is considered in the analysis.

In all, 793 participants mentioned 15,593 alters in the name generator: there were 19.8 contacts on average, with a minimum of 0 and a maximum of 40. In general, respondents who wanted to report more than 40 alters were asked to write these additional names on an extra sheet of paper. However, only a few participants took this opportunity. Avoiding bias from the few egos with more than 40 alters this number is used as a threshold in this section. When analysing the population-wide network all alters are considered.

An overview on the number of contacts is given in Figure 4.8. It shows that a substantial share of the egos has to be excluded from the analysis of network topologies. 49 egos (6.2%) did not report any alters at all. The exclusion is necessary for two reasons: The survey instrument does not allow us to distinguish between people who have no social contacts and those who do not mention them. Completely isolated persons cannot by definition, other than at ego-seed level, be part of a snowball sample. In addition, a minority of 5 egos (0.6%) mentioned only one social contact, which means they were not able to report alter-alter ties in the sociogram. Finally, another 100 egos (13.5%) reported more than one alter but did not report any alter – alter relations in the sociogram. Unfortunately, the questionnaire's design does not allow us to judge whether missing alter – alter relations are due to sparse network structures or item non-response behaviour. On the one side, it is possible that the sociogram data have fatigue effects as they were the last part of the questionnaire. On the other side, it is also possible that these 100 networks are sparsely interconnected, especially as they are of substantially smaller size: There are on average 15.8 social contacts in networks without sociogram information compared to 21.9 alters in those with alter-alter ties. Testing the difference with a t-test shows the result as being significant (p-value = 0.000). Still, it is possible that these structures would introduce bias. In all, these 154 (19.4%) networks are excluded from the descriptive analysis of personal leisure networks topologies. In addition, Figure 4.8 shows a substantial share of egos reporting 40 or more names. However, only the first 40 contacts are considered for analysis. Even though respondents who wanted to report more than 40 contacts were asked to write them down on an extra sheet of paper, it can be assumed that at least some respondents refused to make this additional effort. Therefore, only the first 40 names are considered for analysis of personal networks.

639 egocentric structures (80.6%) are analysed in terms of their topology. Some descriptive statistics are presented in Table 4.7. An average personal leisure network includes 21.9 alters which are connected to each other by 44.8 ties. However, 7.4 (33.8%) alters are reported as isolates. An average network includes 4.1 cliques, where each clique has 4.4 members on average. Some respondents reported the maximum number of 20 cliques. The clique sizes show some heterogeneity. Usually, some cliques are combined into larger components as there are only 2.6 components on average.

'*Density*' describes the proportion of all possible connections between actors of a given graph and all actual connections. The 0.2 average density (which is also the density of the example in Figure 4.7c), including isolates, means that 20% of all possible connections are realized. The proportion of realized connections nearly doubles to slightly over 0.4 when isolates are excluded. Clearly, density values depend on the number of nodes and their average '*degree*' (the degree value informs us of the number of direct ties a node has) in a network. Especially in terms of social networks it is less likely to observe a fully connected graph when many nodes are involved than if only few nodes were present. Therefore, density values and network sizes should be considered together.

Measurements of degree- and betweenness centralization are complementary to the density concept. Both are global measurements for a personal network topology. The concept of '*degree centralization*' defines to what extent a network structure is organized around its most central person. To answer this question the variability of the individual node centrality-indices is quantified. So the first step is to calculate each node's centrality degree, meaning the number of connections each node has to each other node in the graph. A node is central if it has a larger number of connections to other nodes then these other nodes have themselves. The centralization of the entire graph is the differences between the centrality score(s) of the most central node(s) and all other nodes. The centralization index then equals the ratio of the sum of actual differences and the sum of the maximum number of possible differences. A centralization index of 1 results in a star graph like the one in Figure 4.7a (that only includes information from the name generator). A centralization index of 0 results from a graph in which each node is equally connected to each other such in a complete graph (the degree centralization index for Figure 4.7d is 0.3). We decided to calculate degree centralization by excluding isolates, as they cannot hold central positions. Density and degree centralization are complementary measurements as the first concept describes the level of cohesion in a network and the second measures the degree to which this cohesion is compounded by a particular node. With a degree centralization of 0.2 on average, alters in egocentric network components are nearly equally connected to each other, meaning they are usually not organized around a central person. Although there are very centralized networks, up to a degree centralization of 0.8, they are a small minority as the distribution is right skewed.

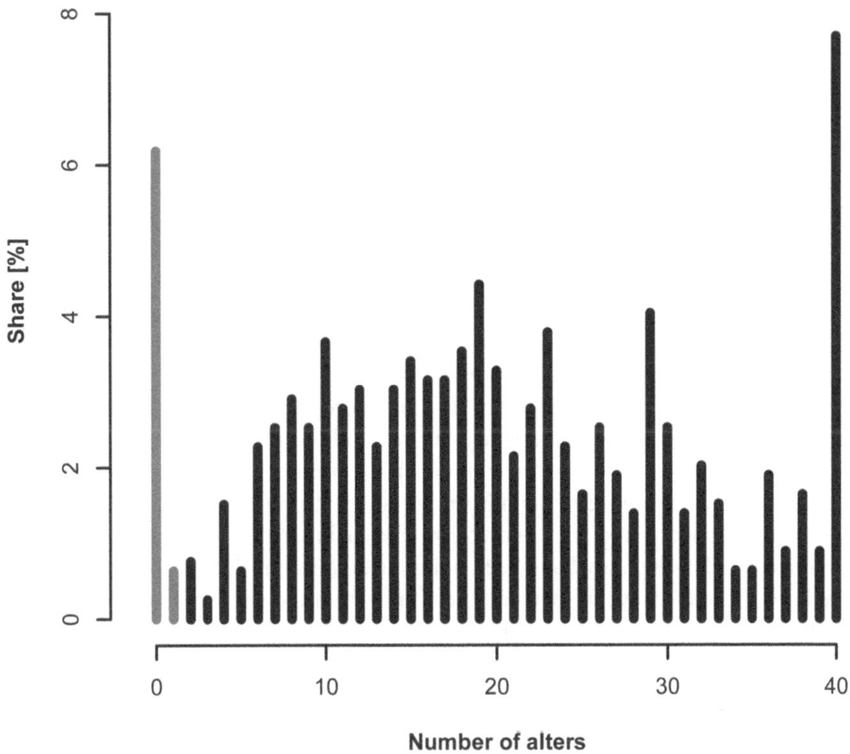

Figure 4.8 Number of reported social contacts (n = 793 egos)

Actor betweenness measures to which extent one node lies on the shortest paths between any two other nodes and thus holds an important position within a network structure. If a node lies 'between' a path connecting other nodes it might have some control over the interaction of these actors. In other words, the interactions of two actors that are not adjacent but connected indirectly might depend to some extent on the behaviour of the node(s) lying between them. In an extreme case, when e.g. a node connects two cliques to one component without any other ties between the two sub-populations, this node possibly controls information flows between the cliques (Trappmann et al. 2011). Actor betweenness is calculated as the number of times a particular node lies on the shortest path between two other nodes, the geodesics, divided by the total number of geodesics connecting these two nodes. So, actor betweenness calculates the probability that two nodes choose a channel that includes the between-node under investigation to pass on whatever is circulating in the network. The global index of '*betweenness centralization*' is standardized and allows comparisons between local nodes in a network in terms of heterogeneity of their betweenness indices. It equals 0 when all nodes have the same local betweenness index and 1 if an actor lies on each possible path between other actors. Again, betweenness centralization is calculated without isolates (Figure 4.7d has a betweenness centralization index of 0.1). Even though both measurements of centralization seem to be very similar, the 'degree-based measure of graph centralization (...) seems to be particularly sensitive to the local dominance of points, while a betweenness-based measure is rather more sensitive to the 'chaining' of points' (Scott 2007: 90; also see Freeman 1979; Wasserman and Faust 2007). The average betweenness centralization of 0.1 suggests that usually all nodes in the egocentric structures are closely connected in the same way, usually without central gatekeepers that connect many alters to each other (for more details on the personal network structures see Kowald and Axhausen 2012a and Kowald and Axhausen 2012b).

Network topology gives information about the arrangement of nodes and ties in a network structure. In addition, it is of interest to focus on nodes' and ties' characteristics. Node characteristics can be studied in terms of homophily, the similarity between egos and their alters. Furthermore, tie strength can be used to distinguish between emotionally important and unimportant contacts for the egos. Both concepts are well known for influencing actors' behaviour. However, in transport planning they have been marginal so far. Introducing some theoretical background and analysing personal networks from the snowball survey in terms of homophily and tie strength aims to show their potentials and identify their influences in approaching and explaining leisure (travel) patterns.

Homophily depends on the characteristic under investigation. Usually, it increases when all kin contacts are excluded, as the ego cannot choose those contacts. In the snowball study 27.9% of the alters are labelled as kin-contacts and 72.1% as non-kin. On the one hand, homophily is an important factor influencing the probability of maintaining contacts and establishing new relationships. In a study of friendship networks among students, the probability of tie formation was much higher when people were similar (Kossinets and Watts 2009). On the other hand, '*status homophily*', resulting from similar characteristics, is also related to '*value homophily*', a term describing similar kinds of behaviour (Steglich and Snijders 2010; McPherson et al. 2001). Therefore, behavioural patterns resulting from homophily are often associated with direct influences. When observing others that successfully satisfy their needs, humans often use copying strategies to achieve the same result. This strategy often makes more sense the more similar archetype and copier are in personal characteristics.

A higher degree of homophily is observed in characteristics over which egos have potential control when choosing others as leisure contacts. Matching this logic, the strongest degree of homophily is observed by age excluding all kin contacts: 80.9% of alters are within ± 10 years of the ego's age; 61.1% of alters match the ego's age ± 5 years. The next significant attribute is sex, with 75.1% of all non-relative alters sharing egos' characteristics. The degree of homophily decreases for education level and civil status and the differences between including or excluding kin contacts also diminish. But similarity is still significant when employing a chi^2–independence test to check expected values against those observed.

Table 4.8 shows some ego – alter attributes checked for similarity. Homophily values were calculated for each personal network separately so that they are independent from network sizes. All egos were included in the analysis. Since the definition of categories influences the degree of homophily they have to be mentioned: Sex is dichotomous. Age similarity is defined as egos' age plus/minus 10 years (for the chi^2–independence test we used five age classes: 0 – 20; 21 – 40; 41 – 60; 61 – 80; 81+). Civil status uses five categories: single, married, divorced, widowed, and living separately. Education is categorized in three classes: primary (elementary school), secondary (high school), and higher education (any kind of university degree). The chi^2–independence test was performed over all ego – alter ties (not for each single personal network structure). Where test results are marked with a flag they might be incorrect because there were too few observations in some cells of the cross-tabulation.

Table 4.8 Homophily-values in personal leisure networks

Attribute	Kind of contact	Homophily [%]	Chi^2-value (p-value)
Sex	All contacts	68.9	1488.2 (0.000)
	Relatives	55.2	7.9 (0.005)
	Non-relatives	75.1	1931.1 (0.000)
Age (+/- 10 years)	All contacts	71.1	3167.5 (0.000)*
	Relatives	51.5	4930.6 (0.000)*
	Non-relatives	80.9	139.6 (0.000)*
Educational level	All contacts	62.3	1666.8 (0.000)
	Relatives	56.1	193.9 (0.000)
	Non-relatives	64.9	1674.0 (0.000)
Civil status	All contacts	60.8	1833.9 (0.000)*
	Relatives	55.6	139.6 (0.000)*
	Non-relatives	62.8	2029.8 (0.000)*

Note: Results with flag (*) might be incorrect because of too few observations in some cells.

The name interpreter asks if respondents would contact their alters when they needed help or wanted to discuss important problems. In the cases egos shared both concerns with an alter the relation was considered as 'strong', otherwise as 'weak'. Considering all ego-alter relations 69.0% were labelled as '*weak ties*' and

31.0% as '*strong ties*'. In addition to homophily, tie strength is crucial in social network analysis as both kinds of ties have different impacts for an ego. Although there are different definitions for tie strength (for an overview see Wellman and Wortley 1990) they all mention similar characteristics:

In terms of relational distance strong ties are closely related to each other. They meet frequently, exchange ideas, and want their feelings to be congruent. Strong contacts are emotionally important for egos as they are a source of support and help in crises. These contacts are also called an ego's core discussion network (Marsden 1987). The stronger the relationship between two persons, the larger is their emotional support for, and their mutual influence on each other (McPherson et al. 2006). Egos and their strong contacts are likely to be densely interconnected and form a group, whereby news, information, or ideas circulate in this group. It is often assumed that relational distance is connected to social distance. Understanding social distance as the similarity between people in terms of socio-demographics or ways of behaviour, it is observed that persons with similar characteristics are often strong contacts. So, homophily and tie strength are complementary. In terms of strong contacts that means that associates with similar characteristics often offer access to similar kinds of resources.

In the case of weak ties the need for cognitive balance is less crucial. Persons, that are weakly connected, meet less frequently, are less emotionally important to each other, and are often sparsely interconnected within an ego's network. Also, the social distance between them is often higher. In terms of socio-demographics and behaviour they are less similar to the ego and one another. Because of their dissimilarity they potentially offer access to a wider range of resources than strongly related persons. Weak ties '*bridge*' egos to persons who are different in terms of personal characteristics, news, information, or ideas. A (local) bridge is a tie that (indirectly) connects many persons to each other and is a crucial part of the shortest paths between them. Thus, weak ties are of importance on both the microscopic as well as the macroscopic level. They link the ego to different spheres of society and transmit information from beyond the egos' personal network horizon. In addition, weak ties are of importance for macroscopic network structures as they connect between different groups. Removing an average weak tie from a global network potentially results in more problems than removing an average strong tie (see Granovetter 1973). So, weak connections are of importance for a society's coherence.

In summary, persons with similarities in socio-demographics and behaviour are likely to form ties and flock together. They are often strongly connected and offer emotional support for each other. For others outside their network they might be less attractive as they themselves are short of fresh information. The advantage of emotional shelter is counterbalanced by low information diffusion rates. The more socio-demographic heterogeneity an ego is connected to, the better this ego is linked to different spheres of society, which is important for e.g. information gathering. These people might be able to hold large numbers of weak connections and thus have access to large amounts of fresh information.

As a disadvantage, persons with only few or no strong contacts have the risk of a lack of emotional support and altruistic help. In a balanced network structure the ego would be linked to both kinds of contacts and have access to both kinds of support: emotional shelter and fresh information. Although the concept of tie strength is clearly related to the theory of 'social capital', the latter is not employed here, as it is notoriously inconsistent in its definition (for an overview on different definitions and comparisons between them see Portes 1998; Lin 1999; Burt 2000; Sobel 2002).

In Otnes (2009) critique on Granovetters work one of his major points concerns the segregation of ties into strong and weak with the help of proxies as not all methods are reliable. Checking this argument we first present some information on the reliability of the survey project's proxy questions: Focussing on the median of the egocentric leisure networks, there are 5 strong and 11 weak social contacts in the average network structure. These numbers are comparable to those mentioned in the literature: In terms of active network sizes, defined as personal relationships over the last two years, Roberts et al. (2009) found the number of emotionally close alters to be 5 and the size of an egos' sympathy group to be between 12 and 15. Concerning the core discussion network of Americans, Marsden (1987) reports an average network size of 3, a number which decreased in the following years (McPherson et al. 2006). Providing a network-based explanation for language, to replace grooming and to allow a larger number of emotionally close contacts, Dunbar (2003) estimated the number of a core discussion network to be around 4, including the ego. The numbers of the present survey match these results. Furthermore, literature on tie strength reports the nuclear family as often being strongly related, whereas weak relations often imply non-kin contacts (see Marsden 1987; McPherson et al. 2006; Roberts et al. 2009). Again, this relation is confirmed by the snowball data, were 69.3% of all nuclear family contacts are labelled as strong, whereas only 23.5% of the remaining contacts are strongly connected. These results support the conclusion that the surveys' two proxy questions for tie strength worked well.

Investigating egos' connectedness to different spheres of society is a proxy for the availability of both kinds of support. Unfortunately, the snowball sample offers only weak indicators for access to social spheres: Homophily or heterophily in terms of sex, age, civil status, and educational level. Studying the effect of tie strength on personal leisure networks is the aim of calculating degrees of status homophily for each network, either for weak and strong contacts twice: once by including ties between egos and their relatives and the second time by excluding all kin contacts.

Results are shown in Figure 4.9. It demonstrates the distinction between relatives and non-relatives as important. Whilst relatives are often strongly connected, the graph clearly shows that they are a substantial source of heterogeneity in both kinds of ties: strong and weak ones. Excluding kin contacts shows character similarity increasing rapidly for strong and moderately for weak ties. This shows again that homogeneity is stronger in contacts that are

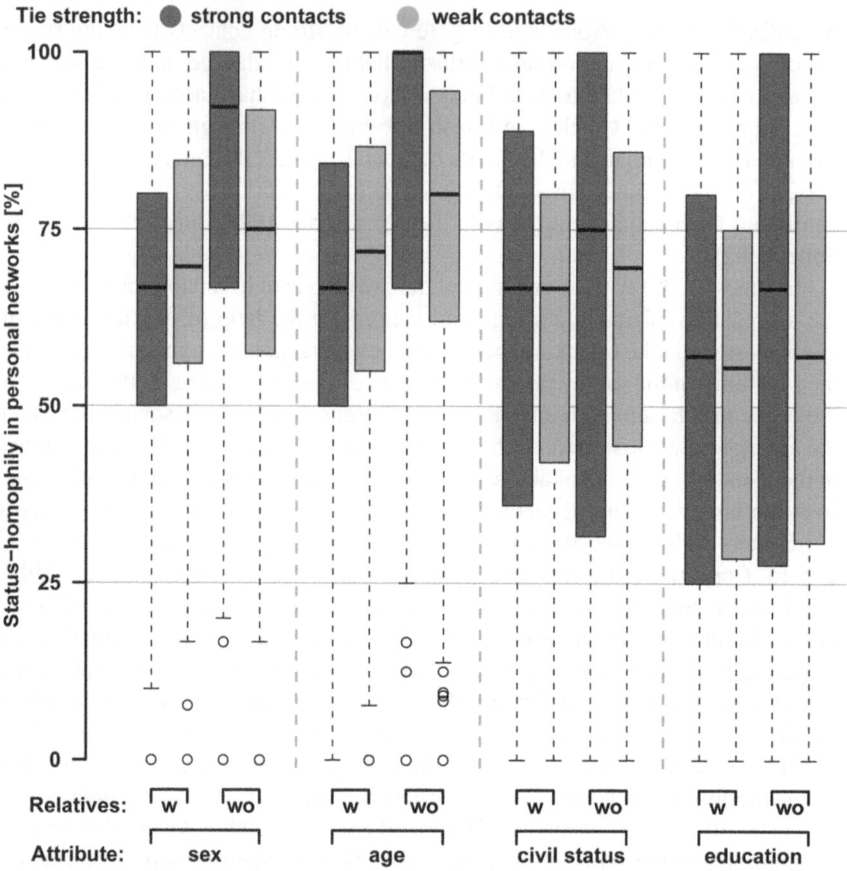

Figure 4.9 Degrees of status similarity in strong and weak contacts (w = with; wo = without)

chosen whilst contact to relatives seems to be maintained independent from characteristic similarity. When relatives are excluded, weak ties characteristics are more heterogeneous and potentially offer more access to different social spheres than strong tie contacts. When relatives are included, the differences in strong- and weak ties heterogeneity are rather small. So, weak ties might bridge egos to different spheres of society but relatives might, even when they are frequently labelled as strong contacts, too. On the other side, it has to be mentioned, that heterogeneity for relatives in some characteristics usually comes together with high amounts of homophily in other characteristics, e.g. religion, race and ethnicity (see McPherson et al. 2001), which is not necessarily the case for non-kin contacts.

As mentioned before, the sample, resulting from a snowball chain of participants, who could live anywhere in the world, is highly clustered in Switzerland. Within Switzerland, most persons live in the German speaking area, and, more specifically, in Canton Zurich. Over all ego – alter ties the average distance is 107 km and the median around 9 km. Distances are calculated as great circle distances (crow fly distances) employing egos' home locations as the base. The distribution is heavily right skewed with a minimum of 0 metres, in the case of people sharing the same apartment, and 17'900 km at maximum. Most contacts live somewhat close to their egos, usually within a 50 kilometres radius around the egos' home location.

Figure 4.10 summarizes the shares of leisure contacts living at certain distance classes. It also includes information on tie strength and the share of relatives among strong contacts. Distance class 1 can be labelled as local contacts living less than 1 km away from the egos' home location. It contains nearly 18% of all leisure contacts. These contacts can be reached by foot and are labelled as '*everyday-life contacts*'. This confirms findings reported by Wellman (1990), which showed nearly 20% of all active ties to be persons from respondents' neighbourhoods. Ties in this class are often weakly connected to their ego as, overall, the share of strong contacts is lower when compared to the other distance classes. It is remarkable that only few relatives live in this distance class, but that they are often mentioned as strong contacts (Kowald and Axhausen 2011).

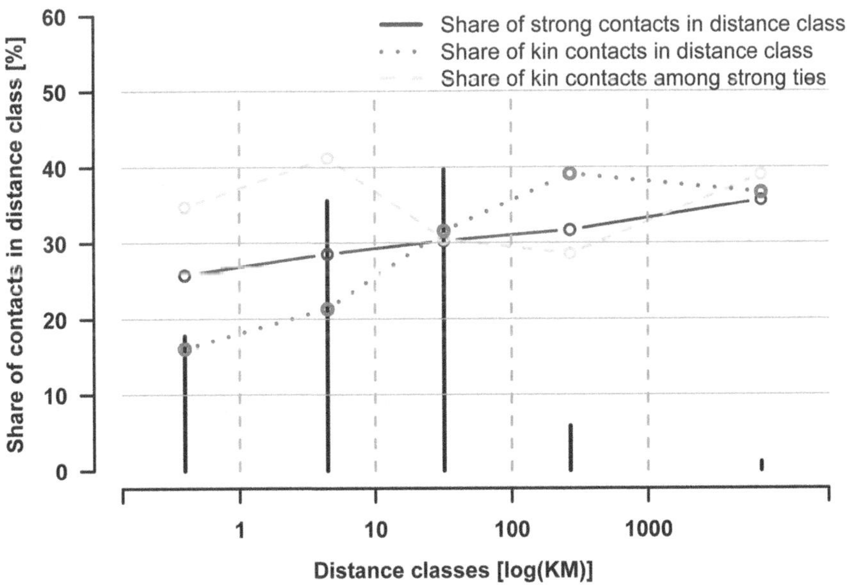

Figure 4.10 Distance classes and tie strength

36% of all alters live in the second distance class, between 1 and 10 km away. They can be labelled as '*regional contacts*'. Still, they live within walking distance and probably in the same municipality as their egos but one might prefer mobility tools for meetings. The share of strong contacts increases to nearly 29%. Also the share of relatives (21%) is larger in comparison to distance class one. Similarly to the former distance class relatives are often strong contacts.

Distance class three, from 10 to 100 km, can, at least in the Swiss context, be labelled '*nationwide relationships*' and contains around 40% of the alters. Meetings with these persons certainly include the use of mobility tools like public transport or private vehicles. If these tools are available distances of up to 100 km are still practicable in daily life. The share of strong contacts is around 30%, which is similar to the share of relatives in the distance class and the share of relatives within strong contacts.

Only around 6% of all leisure contacts are '*European-wide*', living at distances between 100 and 1,000 km. Again we find an increased share of strong contacts of 32% when compared to the former distance class. In contrast to the former classes the relation between share of relatives and the number of relatives used as strong ties changes: Whilst there are many relatives, 39% of all contacts, they are rarely strong contacts.

Finally the picture is completely different for the last distance class, which contains only 1% of all alters, the '*international ties*', living more than 1,000 km away. Overall, this class shows the highest shares of strong contacts (36%). In addition, the share of relatives within social contacts is still high (37%) and these relatives are often used as strong contacts (39%).

In summary, respondents mix local, nationwide and international ties in their personal network, although most leisure and emotionally important contacts seem to live near the egos' home locations. This finding is confirmed by other studies (Larsen et al. 2006; Carrasco 2006; also see Chapter 3), and highlights the importance of persons living in the same urban or rural area. An increasing geographical distance limits the opportunities for meeting each other and removes the possibilities for certain kinds of support, which can be important in daily life (see Wellman and Wortley 1990). The share of strong contacts increases with an increasing distance. This is related to relationship duration, which also increases for each distance class. Whilst ties in distance class one are on average 18 years old, relationships in distance class five are 27 years old. The correlation between distance and relationship duration is low but highly significant: A Spearman's rank correlation test shows a correlation of 0.105 (p-value = 0.000) when the 0.95-quantile is used as a cut-off point to exclude the most extreme values in both relation duration and distance. So, the further away the alter live the higher the chances that the ego is engaged with them for emotionally important matters.

The low share of strong contacts within the first two distance classes do not match the theory of tie strength well. It suggests strong contacts meeting frequently and influencing each other. However, contacts at small distances usually meet frequently too. In addition, there are good reasons to assume that near-by contacts, whether weakly or strongly connected to the ego, are influential in terms of behaviour. Copying behaviour is a well-known human strategy for need satisfaction (see Aars et al. 2004; Junker 2006). Although the literature on contagion behaviour highlights the influence of strong contacts there are several other influences. For example people moving in a messy environment tend to start littering their trash instead of putting it into rubbish bins (Cialdidni et al. 1990) and suicide rates increase when people see, read, or hear that others with similar socio-demographics committed suicide (Phillips 1979). In addition, especially when quick decisions are needed people focus on others whose behaviour can be observed directly (for emergency evacuations see Quarantelli 1954). In all these situations people are influenced neither by strongly nor by weakly connected social contacts but by completely unknown others. They might, at least in part, be unaware of these influences, as they are by definition hard to capture in survey studies. Nevertheless, they exist and from this perspective the concept of tie strength only explains a part of the story of how people are influenced by others. Of course, a strong tie that lives close by might be more influential than a weak contact living close by, but still, a weak contact seen daily certainly has some influence, which, at least in terms of contagion behaviour, might be more important than the influence of a strong contact living 10'000 km away. In conclusion, it seems worth extending the theory of tie strength with spatial considerations.

In summary, the survey instrument of the snowball study successfully collected information on respondents' personal networks. Analysis results on 639 egocentric network structures showed that an average network of leisure and emotionally important people includes 22 alters. Extending information from the name generator with data from the sociogram identifies seven of these alters as isolates, persons who are only in contact with ego but do know any other alter in the network. The remaining 15 alters participate in four leisure cliques with an average of four members per clique. However, people in the networks are usually sparsely interconnected and there are no central actors in powerful positions.

An analyses of egos and alters socio-demographic similarity shows higher homophily values for kin contacts compared to non-kin contacts. In addition, similarities which can be chosen arbitrarily and do not change over the life-course (e.g. sex and age) have higher homophily values. Characteristics which might change over the life time (e.g. educational level and civil status) show increased homophily values as well, especially when relatives are excluded. However, related homophily values are lower than those for non-changing characteristics.

Kin contacts are a source of socio-demographical heterogeneity. In addition, they are often strongly connected with their ego. These results do not fit the theory on tie strength well. Whilst weak ties are known for being different (in behaviour and socio-demographics) and bridging an ego to different spheres of society our findings show that strongly connected kin contacts differ from their egos in terms of socio-demographics and might be able to build such a bridge as well. In addition, it would be worth extending the theory on tie strength with spatial issues to avoid conflicts with the well-documented phenomenon of contagious behaviour:

Weakly connected social contacts living nearby might be more influential than strong contacts who live further away as their behaviour can be observed directly in everyday life.

Empirical Results: The Population-wide Network Structure

Information on connected egocentric networks or even a population-wide (leisure) network structure is needed when aiming to impute a social network in agent-based simulations. The influence and importance of personal networks' interconnectivity can easily be understood when recalling one basic finding from Chapter 3: People only have limited resources for maintaining direct contacts. However, usually they are connected to large numbers of people beyond their personal networks' horizon in an indirect way. In everyday life these indirectly connected people are often called 'friends of a friend' or labelled by similar terms depending on the kind of relationship and the order of interconnectivity (if they are one or more intermediate node away). This interconnectivity is a source of important influences that have to be considered in a network generation approach. It is the global network structure that allows mental and material artefacts to circulate within any community. Examples for circulating artefacts are feelings (Christakis and Fowler 2009), styles and behaviours (Gladwell 2002), as well as contagious diseases (Smieszek et al. 2011). At least some of these shared artefacts facilitate a joint group identity and are thus a source for social coherence (Scott 2007; Mewes 2009; Trappmann et al. 2011). Knowledge about this population wide structure is crucial to implement a leisure network in disaggregate activity-based models and to allow a consideration of contacts' influence on individual activity travel behaviour.

Employing a snowball sample to survey a population wide network means that two kinds of overlaps between the embedded egocentric network structures are possible. At first, overlaps occur due to the sampling strategy: Egos report their alters and some of these alters become egos in the process of recruitment themselves. These personal network overlaps result in direct ego – ego ties.

The second source of overlaps connects components of connected personal networks to each other. These '*component-overlaps*' occur when an ego from one component mentions an ego or an alter from any other component as a social contact. So, these overlaps can either be direct ego – ego ties or ego – alter – ego link chains. Whatever their form, they connect two formerly separated components of connected personal networks into one larger '*giant-component*'.

An empirical example for both kinds of overlaps is given in Figure 4.11. Figure 4.11a shows a network structure including two seeds, and several other egos and their alters. A clearer representation of the component's structure can be achieved when excluding all nodes that are connected to only one other node (as done in Figure 4.11b). What remains is a graphical representation of all nodes, which are connected in multiple ways. These nodes play a key role for the giant-component as they interconnect both egocentric network structures and formerly separated components of connected personal networks.

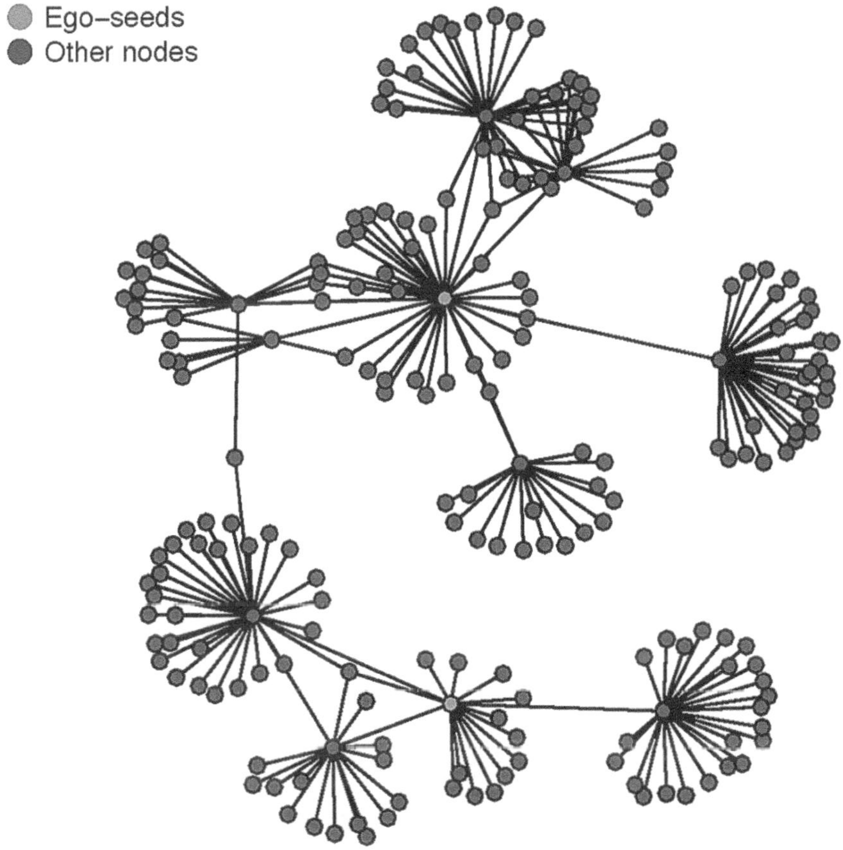

Figure 4.11a Analysis of the global snowball network (component overlap: all nodes)

Positions in the network structure are analysed to detect those nodes that potentially control interactions and information flows between other nodes. Figure 4.11b includes three nodes in important positions: Two of them are seeds, one is an alter that connects both seeds in an indirect way. Each of these three nodes is a '*cutpoint*'. If one or several of them were removed from the graph it would split up into several smaller components. So, a cutpoint is an important node as it bridges (otherwise separately) sub-graphs (Trappmann et al. 2011, Wasserman and Faust 2007; Scott 2007). A cutpoint's importance for the component can change in the course of the further data collection. If additional ties between sparsely connected sub-graphs are found on the next iteration level the importance of a cutpoint decreases and it becomes one connecting node amongst others. The structure would not split up any more if it was removed.

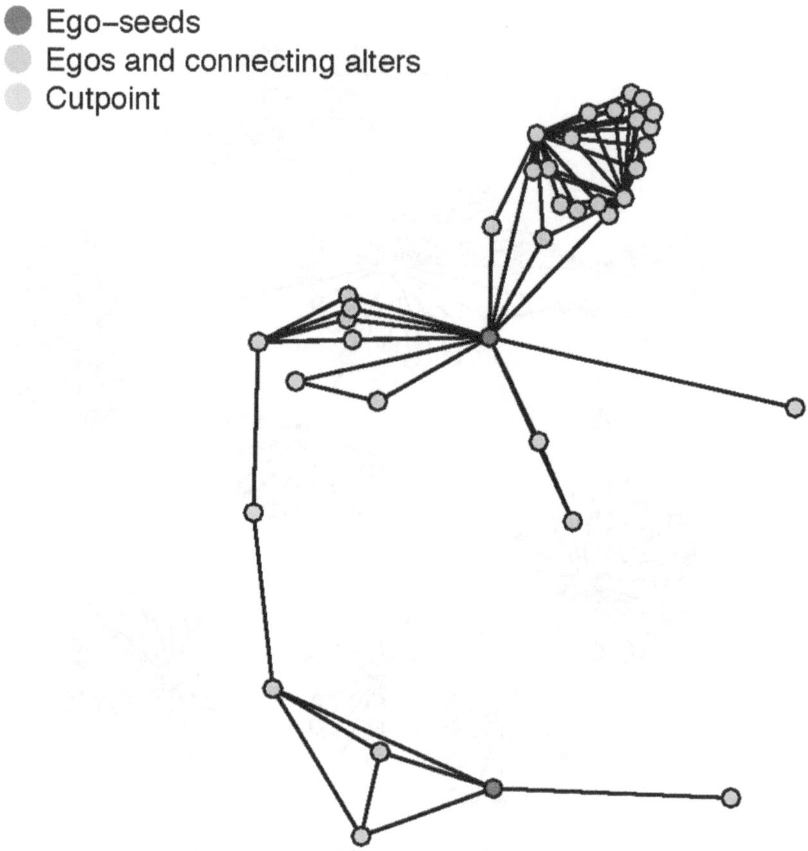

Figure 4.11b Analysis of the global snowball network (component overlap: only connecting nodes)

Before focussing on descriptive statistics from the snowball graph in detail it has to be mentioned that only the first three iterations are analysed separately. The snowball sample was stopped on the fourth iteration after only a few egos had responded (as mentioned above). Therefore, the fourth iteration is not analysed separately. However, fourth iteration egos and their alters are treated as being a part of iteration three and are thus included in the analysis. Furthermore, the snowball graph includes information from 22 egos that mentioned more than 40 alters. These additional 361 alters in total were excluded from the analysis of isolated personal networks to avoid bias (because the name generator only provided space for 40 names). However, in the analysis of the entire snowball graph they can make a substantial difference in the form of a component-overlap, for which they are included.

Descriptive statistics on the snowball graph clearly show its connectedness becoming denser with each iteration step: More and more formerly separated components overlap. As already mentioned above, the sample started with 38 isolated components on the ego-seed level. The smallest component includes one seed that mentioned only one alter. Already on iteration 1 components begin to overlap. Two giant-components with each of them including two seeds and one additional cutpoint are observed. The largest of these giant-components includes 215 nodes. The smallest component still includes one seed and one alter, and is thus an example of a snowball component which came to an end: The single alter refused participation. On iteration 2 there are three giant components including 10, 3, and 2 seeds. The total number of components thus decreases to 26. Those components that include only one seed are of small size whereas the ten seeds giant-component includes 3,750 nodes. Finally, on iteration 3 there are only 18 components left: 17 of them include only one seed. In addition, there is one giant-component that includes 21 seeds, three cutpoints, and 12,922 nodes in total.

Table 4.9 Overlaps between personal networks

	Seeds	Iteration 1	Iteration 2	Iteration 3
Number of Components	38	36	26	18
Component size [No. of nodes]				
Minimum	2.00	2.00	2.00	2.00
1st-quantile	9.00	25.00	14.50	12.25
Median	14.00	52.00	89.00	38.00
3rd-quantile	20.75	73.25	192.80	100.50
Maximum	39.00	215.00	3,750	12,922
Mean	15.95	61.33	269.50	772.40
Component overlaps				
No. of giant-components	–	2	3	1
No. of cutpoints	–	2	5	3
Connected seeds in smallest giant-component	–	2	2	21
Connected seeds in largest giant-component	–	2	10	21
Connected seeds total	–	4	15	21

Results show that the snowball method was successful. Although there are components that stopped on iteration 1, many others went on through the recruitment process. Components did not only result in connected personal networks but also started to overlap and form giant-components. This is influenced by the decision to start snowball chains in Canton Zurich, on the one side, and the geographically clustered structure of the leisure contacts, on the other side. The fact that there is only one giant-component including the majority of seeds left on iteration 3 and a minority of isolated components including only one seeds suggest that further component overlaps could have been expected on iteration 4. Unfortunately the snowball was stopped at this point.

Six degrees of separation is the result of Stanley Milgram's experiment on the nature of the 'small world effect' (Milgram 1967). This effect describes the amazement of two people meeting by chance and recognizing that they share a social contact (which is often expressed by a surprised statement like 'we live in a small world').

In the original experiment some initial US respondents were asked to forward a package to a predefined recipient. This recipient was chosen randomly from the US population. Senders were told the recipient's name and address. If the recipient was unknown to the sender she was asked to forward the package to a person that she knows on a first name base and of whom she thought that he could either be known to the recipient personally or that he could be in a good position to forward the package to a person that might know the recipient (see Milgram 1967, Travers and Milgram 1969).

Only considering paths that reached the recipients, Milgram found a median of five intermediate persons between sender and recipient, which, in other words, means six ties between them. This and similar results, the six degrees of separation, was repeatedly reported in a number of studies since then (for a global experiment with e-mail messages see Dodds et al. 2003; for an overview on small world experiments see Kochen 1989; Schnettler 2009a; for a critical review see Kleinberg 2006). Employing Facebook data and thus calculating the shortest path between any two randomly drawn nodes instead of analysing the routed path of a forwarded message resulted in an even lower degree of separation: 4.74 ties are necessary on average (Backstrom et al. 2011; for an overview on decentralized shortest paths search algorithms see Kleinberg 2006).

The shortest paths in the snowball data, also labelled geodesic distances or, in case of a small world analysis, degrees of separation, are calculated between all nodes and exclusively between seeds. This seems reasonable as seeds were recruited from a random sample. Connections between seeds can thus be interpreted as connections between randomly chosen persons. Considering any pair of seeds and staying with the concept of Milgram's experiment, these persons are both the initial senders and the pre-defined recipients of a message.

Focussing on all nodes on the seeds' level there is a minimum distance of one tie, an ego-alter relationship, and a maximum distance of two ties, which is an alter-ego-alter link chain. Longer chains are not possible because there are no giant-components on the ego seeds' level (see Table 4.10). On iteration 1 there are such overlaps and paths length thus increases. The shortest path still includes only one, the longest path includes eight intermediate ties. The average path length between all nodes is around four ties. As there are giant-components on iteration 1, geodesic seed to seed distances can be calculated. On iteration 1 the geodesic distance between seeds in giant-components includes four intermediate ties. Average paths length further increases on iteration 2: It is around eleven intermediate ties when all nodes are considered, with a maximum of 24 ties, and around seven degrees of separation on average for seed to seed connections, with a maximum of 18 intermediate ties. Finally, paths lengths between all nodes are similar on iteration 3: The shortest path on average is around eleven intermediate ties. In terms of seed to seed connections it further increases and is around twelve ties.

Table 4.10 Geodesic distances between nodes

	Seeds	Iteration 1	Iteration 2	Iteration 3
All nodes				
Minimum	1.00	1.00	1.00	1.00
1st-quantile	2.00	2.00	7.00	9.00
Median	2.00	3.00	10.00	11.00
Mean	1.90	3.56	10.68	11.47
3rd-quantile	2.00	4.00	14.00	14.00
Maximum	2.00	8.00	24.00	27.00
Seed to seed connections				
Minimum	–	4.00	4.00	4.00
1st-quantile	–	4.00	5.67	10.00
Median	–	4.00	7.00	12.50
Mean	–	4.00	7.36	11.97
3rd-quantile	–	4.00	9.08	14.00
Maximum	–	4.00	18.00	21.00

These results do not confirm the six degrees of separation rule for the snowball data. However, it has to be mentioned that the results are not sufficient enough to conclude that there are more intermediate ties between any randomly chosen pair of persons from the Swiss leisure network. Results are clearly influenced by the snowball methodology: Paths between seeds are long when other seeds and resulting components of connected egocentric networks lie between them. It can be assumed, that most paths between seeds in a giant-component, like the one on iteration 3 that includes 21 seeds), have to cross intermediate seeds and associated network structures. In such cases it is unclear if the empirically calculated shortest path is identical to the shortest path in reality or if an important bridge between two components is missing. The increasing path length between seeds is a sign that the snowball was stopped too early to allow detailed analysis on this question. Furthermore, it has to be kept in mind that Milgram's participants were allowed to employ any type of contact they knew whereas the snowball sample only includes leisure and emotionally important contacts.

An overview on the global statistical properties of the iteration 3 giant-component is given in Table 4.11 for reasons of completeness. These figures are helpful when trying to compare the data to results from other survey efforts on population-wide networks. However, information on e.g. betweenness centrality are sensitive to shortcuts in a network structure, which should be considered in such comparisons.

Table 4.11 Global characteristics of the largest giant-component

Vertices	Edges	Density	Transitivity	Closeness	Betweenness	Centralization
12,922	14,573	0.000	0.016	0.111	0.294	0.012

In summary, the snowball sample successfully collected information on connected personal networks. More than that, it also collected information on overlaps between components of personal networks. However, at the time when most components were already connected to each other the sample was stopped. There are good reasons to assume that taking the effort of one or two further iterations would have been worthwhile and resulted in a more detailed and realistic picture of the Swiss-wide leisure network. Whether this extended data set would have corresponded with observation results from other studies on the six degrees of separation cannot be judged. For sure, additional ties and shortcuts between nodes would have increased the reliability of the results.

Conclusions

The snowball study employed a survey method and instrument that allows the collection of information on both personal networks as well as a population-wide network structure. This program contributes to the scientific discussion in two dimensions: There are empirical results on network structures and there are results on survey methods.

A combination of the personal network approach and snowball sampling was chosen as the best fitting survey method. This combination of a sampling and a survey method does not stop recruitment after the initial respondents (like the personal network approach), but continues data collection by asking respondents' social contacts for participation as well. Used in this way, snowball sampling extends the personal network approach and surveys chains of connected egocentric network structures to get an idea about the underlying population-wide network structure.

The survey instrument asked respondents for information on four topics: Egos' characteristics, egos' leisure and emotionally important contacts, these alters' characteristics, and the alter-alter relationships. Concerning the ego-alter relationship it allows a distinction between strongly and weakly related contacts and calculations of homophily values. A pre-test shows that name generators are sensitive even to small changes as the number of reported social contacts changed with the layout of the name generator.

Snowball sampling includes three major sources for bias: Selection-, homophily-, and degree bias. The survey team developed different measures to address bias: A multi-contact strategy, a web-page, a greeting card, and a gratification strategy were employed successfully. These measures helped to create a trustworthy atmosphere and increase the response rate. An unexpected result was found for the effect of the gratification strategy. Significantly more persons than expected, respondents and non-respondents, sent back the appreciation. Recruiting 40 seeds to start the snowball chains in two consecutive sub-samples with 20 seeds each helped to balance homophily bias. In addition the project team in Berlin developed a method to balance degree-bias (Illenberger and Flötteröd 2012).

An analysis of the fieldwork shows that both the survey method and the developed survey protocol worked better than expected: Originally it was planned to start the sample with 60 seeds in three consecutive subsamples and recruit around 500 egos within two snowball iterations. Exceeding this expectation around 500 respondents participated on the second iteration of sub-sample 2. In conclusion, there is, at least to our knowledge, no other study of that size, which employs snowball sampling to survey information on a population-wide network without reducing the target population to specific personal characteristics or communication modes.

Analyses on data quality show increased shares of missing values for some items, particularly those with a higher response burden, like difficult recall tasks and unusual designs, that ask for confidential information, like names and postal addresses of associates, and that are asked at the end of the survey instrument. There are clear hints that respondents' motivation is a key element in issues of missing values. Highly motivated respondents report larger network sizes with lower shares of missing values on average.

The analysis of personal networks aims to investigate the structure of complete leisure networks, including alter-alter ties from the sociogram, study the role of homophily and tie strength in leisure networks. Focussing on respondents that provided information on alter-alter relationships in the sociogram exclusively, there are around 640 personal network structures. However, a 23% response rate implies a highly selective response behaviour and is clearly not enough to represent any target population's characteristics sufficiently. In addition, the sample combines snowballs sampling and the name generator approach and thus includes bias.

An average structure includes around 22 alters, which are connected by nearly 45 ties. Most alters within a network are connected to some others in leisure cliques, which include around four members on average. Often some alters are members in more than one clique and thus connect these cliques into components. In all characteristics, there is a lot of variety in personal networks' structure.

Issues of homophily and tie strength are known for influencing peoples' behaviours. Personal networks in the sample show higher shares of homogeneity for characteristics that are usually stable over a life time, like sex and age. In contrast, there are lower shares of similarity for characteristics that might change, like educational level and civil status. Furthermore, higher shares of homogeneity are observed for chosen contacts, whereas relatives, who are often mentioned as leisure contacts, usually have lower shares. In terms of tie strength strong contacts show high shares of homophily, which is especially true when relatives are excluded from the analysis. In contrast, there are higher shares of heterophily for weakly connected associates. So, weakly connected alters have different socio-demographics in comparison to their egos. Employing socio-demographics as a proxy for different spheres of society they might bridge their egos to groups of persons beyond their own horizons. This finding is in line with the theory of tie strength. However, the story is different as soon as relatives are included in the analysis. Then also strongly connected alters show higher shares of heterophily and might be able to bridge their egos.

Extending the personal network approach with snowball sampling, the survey successfully collected information on components of connected personal networks. In addition, these components started to overlap during recruitment. On the final iteration, they form a giant component, which includes the vast majority of all nodes and ties. Although the snowball approach worked better than expected there are clear hints that the sample was stopped too early. One hint is an increasing average shortest paths length between randomly chosen nodes the higher the iteration level under consideration. If the network structure was (already) complete, this average geodesic distance should be stable. So, each new iteration level resulted in substantial differences in the network structure.

However, the surveyed population-wide network structure represents the best chance to study characteristics of the Swiss leisure network, as there is no similar study available. Data are used to extend an existing network generation approach. This approach was originally designed to generate isolated personal networks. In its extended version it is able to generate a population-wide network for a synthetic population (Arentze et al. 2013; Kowald et al. forthcoming).

References

AAPOR (2011) Standard Definitions: Final Dispositions of Case Codes and Outcome Rates for Surveys, *The American Association for Public Opinion Research.*

Aarts, H., P.M. Gollwitzer and R.R. Hassin (2004) Goal contagion: Perceiving is for pursuing, *Journal for Personality and Social Psychology*, 87(1): 23–37.

Arentze, T.A., M. Kowald and K.W. Axhausen (2013) An agent-based random-utility-maximization model to generate social networks with transitivity in geographic space, *Social Networks*, 35(3): 451–9.

Axhausen, K.W., C. Weis (2010) Predicting response rate: A natural experiment, *Survey Practice*, 3/2 (2010) http://surveypractice.org, accessed on April 2010.

Backstrom, L., P. Boldi, M. Rosa, J. Ugander and S. Vigna (2011) Four degrees of separation, http://arxiv.org/abs/1111.4570, accessed December 2012.

Bekhor, S., Y. Cohen, C. Solomon (2013) Evaluating long-distance travel patterns in Israel by tracking cellular phone positions, *Journal of Advanced Transportation*, 47: 435–46.

Bell, D.C., B. Belli-McQueen and A. Haider (2007) Partner naming and forgetting: Recall of network members, *Social Networks*, 29: 279–99.

BFS and ARE (2012) Mobilität in der Schweiz, Ergebnisse des Mikrozensus Mobilität und Verkehr 2010, *Federal Office for Spatial Development and Swiss Federal Statistical Office*, Bern and Neuenburg.

Burt, R.S. (2000) The network structure of social capital, in B. Staw and R. Sutton (eds) *Research in Organizational Behaviour*, New York: JAI Press, pp. 345–423.

Campbell, K.E. and B.A. Lee (1991) Name generators in surveys of social networks, *Social Networks*, 13: 203–21.

Carrasco, J.A. (2006) Social activity-travel behaviour: A personal network approach. Dissertation, University of Toronto, Toronto.

Carrasco, J.A., B. Hogan, B. Wellman and E.J. Miller (2008) Collecting social network data to study social activity-travel behaviour: An egocentric approach, *Environment and Planning B*, 35(6): 961–80.

Chartrand, T.L. and J.A. Bargh (1999) The chameleon effect: The perception-behavior link and social interactions, *Journal of Personality and Social Psychology*, 76(6): 893–910.

Christakis, N.A. and J.H. Fowler (2009) *Connected: The Surprising Power of Our Social Networks and how they Shape Our Lives*, New York: Little Brown.

Cialdini, R.B., R.R. Reno and C.A. Kallgren (1990) A focus theory of normative conduct: Recycling the concept of norms to reduce littering in public places, *Journal of Personality and Social Psychology*, 58(6): 1015–26.

Coleman, J.S. (1958) Relational analysis: The study of social organizations with survey methods, *Human Organization*, 17(4): 28–36.

Couper, M.P. and P.V. Miller (2008) Web survey methods: Introduction, *Public Opinion Quarterly*, 72(5): 831–5.

Diekmann, A. (2001) *Empirische Sozialforschung*. Grundlagen, Methoden, Anwendungen, Rohwolt, Reinbek.

Diekmann, A. and B. Jann (2001) Anreizformen und Ausschöpfquoten bei postalischen Befragungen. Eine Prüfung der Reziprozitätshypothese, *ZUMA-Nachrichten*, 48(25): 18–27.

Dillman, D.A. (2000) *Mail and Internet Surveys: The Tailored Design Method*, New York: Wiley & Sons.

Dodds, P.S., R. Muhamad, and D.J. Watts (2003) An experimental study of search in global social networks, *Science*, 301: 827–9.

Dunbar, R.I.M. (2003) The social brain: Mind, Language, and society in evolutionary perspective, *Annual Review of Anthropology*, 32(1): 163–81.

Erickson, B.H. (1979) Some problems of Inference from chain data, *Sociological Methodology*, 10(1): 276–302.

Fisher, C.S. (1982) *To Dwell Among Friends: Personal Networks in Town and City*, Chicago: University of Chicago Press.

Freeman, L.C. (1979) Centrality in social networks: Conceptual clarification, *Social Networks*, 1(1): 215–39.

Frey, B.S. (1997) *Not Just For the Money: An Economic Theory of Personal Motivation*, Cheltenham, UK and Brookfield, VT: Edward Elgar.

Fu, Y.C. (2007) Contact diaries: Building archives of actual and comprehensive personal networks, *Field Methods*, 19(2): 194–217.

Gabler, S. (1992) Schneeballverfahren und verwandte Stichprobendesigns, *ZUMA-Nachrichten*, 16(31): 47–69.

Gladwell, M. (2002) *The Tipping Point: How Little Things Can Make a Big Difference*, New York: Little Brown.

Granovetter, M. (1973) The strength of weak ties, *American Journal of Sociology*, 78(6): 1360–80.

Groves, R.M., F.J Fowler, M.P. Couper, J.M. Lepkowski, E. Singer and R. Tourangeau (2009) *Survey Methodology*, Hoboken, NJ: Wiley.

Heckathorn, D. (1997) Respondent-driven sampling: A new approach to the study of hidden populations, *Social Problems*, 44(2): 174–99.

Heerwegh, D. and G. Loosveldt (2008) Face-to-face versus web surveying in a high-internet-coverage population. Differences in response quality, *Public Opinion Quarterly*, 72(5): 836–46.

Hogan, B.J., J.A. Carrasco and B. Wellman (2007) Visualizing personal networks: Working with participant-aided sociograms, *Field Methods*, 19(2): 116–44.

Illenberger, J. and G. Flötteröd (2012) Estimating properties from snowball sampled networks, *Social Networks* 34(4): 701–11.

Jansen, D. (2003) *Einführung in die Netzwerkanalyse*. Grundlagen, Methoden, Forschungsbeispiele, Opladen: Leske + Budrich.

Jones, E.C. (2003) Class-based social networks in regional economic systems, *Research in Economic Anthropology*, 22(1): 3–23.

Junker, T. (2006) *Die Evolution des Menschen*, München: Beck.

Kadushin, C. (2005) Who benefits from network analysis: Ethics of social network research, *Social Networks*, 27: 139–53.

Kleinberg, J. (2006) Complex networks and decentralized search algorithms, *Proceedings of the International Congress of Mathematics*, Madrid.

Klovdahl, A.S. (2005) Social network research and human subject protection: Towards more effective infectious disease control, *Social Networks*, 27: 119–37.

Knoke, D. and J.H. Kuklinski (1982) *Network Analysis*, Newbury Park, CA: Sage.

Kochen, M. (1989) *The Small World*, Norwood, NJ: Ablex.

Kossinets, G. and D.J. Watts (2009) Origins of homophily in an evolving social network, *American Journal of Sociology*, 115(2): 405–50.

Kowald, M., A. Frei, J.K. Hackney, J. Illenberger and K.W. Axhausen (2009) Using an ascending sampling strategy to survey connected egocentric networks: A field work report on phase one of the survey, *Arbeitsberichte Verkehrs- und Raumplanung*, 582, IVT, ETH Zurich, Zurich.

Kowald, M. and K.W. Axhausen (2011) Strong and weak relationships: Tie strengths in egocentric leisure networks, paper presented at the *106th American Sociological Association Annual Meeting*, Chicago, August 2011.

Kowald, M. and K.W. Axhausen (2012a) Focussing on Connected Personal Networks: Selected Results from a Snowball Sample, *Environment and Planning A* 44(5): 1085–100.

Kowald, M. und K.W. Axhausen (2012b) Egos' horizons and behind it: Snowball sampling of personal leisure networks, in M. Grieco and J. Urry (eds) *Mobilities: New Perspectives on Transport and Society*, Farnham: Ashgate, pp. 109–34.

Kowald, M. (2013) Focussing on leisure travel: The link between spatial mobility, leisure acquaintances and social interactions. *Dissertation*, IVT, ETH Zurich, Zurich.

Kowald, M., P. van den Berg, A. Frei, J.-A. Carrasco, T. Arentze, K.W. Axhausen, D. Mok, H.J.P. Timmermans and B. Wellman (2013) Distance patterns of personal networks in four countries: A comparative study, *Journal of Transport Geography*, 31: 236–48.

Kowald, M. and K.W. Axhausen (2014) Surveying data on connected personal networks, *Travel Behaviour and Society*, 1(2): 57–68.

Kowald, M., T.A. Arentze and K.W. Axhausen (forthcoming) Individuals' spatial social network choice: Model-based analysis of leisure-contact selection, submitted to Environment and Planning B.

Larsen, J., J. Urry and K.W. Axhausen (2006) *Mobilities, Networks, Geographies*, Aldershot: Ashgate.

Lin, N. (1999) Building a network theory of social capital, *Connections*, 22(1): 28–51.

Lindenberg, S. (2006) Prosocial behavior, solidarity, and framing processes, in D. Fetchenhauer, A. Flache, A.P. Buunk and S. Lindenberg (eds) *Solidarity and Prosocial Behavior. An Integration of Sociological and Psychological Perspectives*, Waterloo: Springer, pp. 23–44.

Lindenberg, S. (2008) Social rationality, semi-modularity and goal-framing: What is it all about?, *Analyse & Kritik*, 30(1): 669–87.

Mamali, C. (2006) Contextual pictorial representations of self-others relationships: The social autograph technique, http://www.self.ox.ac.uk/Conferences/2006/Mamali.pdf, August, 1st 2011.

Makoto, C., J. Zhang, A. Fuhiwara (2012) Effects of nursing care on time use behaviour: A comparative analysis in Japan, *Transportmetrica*, 8: 181–208.

Marsden, P.V. (1987) Core discussion networks of Americas, *American Sociological Review*, 52: 122–31.

Marsden, P.V. (1990) Network data and measurement, *Annual Review of Sociology*, 16 435–463.

Mathews C., N. Coetzee, M. Zwarenstein, C. Lombard, S. Guttmacher, A. Oxman and G. Schmid (2001) Strategies for partner notification for sexually transmitted diseases, *Cochrane Database of Systematic Reviews*, 2, Wiley, the Cochrane library.

Matzat, U. and C. Snijders (2010) Does the online collection of ego-centred network data reduce data quality? An experimental comparison, *Social Networks*, 32(2): 105–11.

McPherson, M., L. Smith-Lovin and J.M. Cook (2001) Birds of a feather: Homophily in social networks, *Annual Review of Sociology*, 27: 415–44.

McPherson, M., L. Smith-Lovin and M.E. Brashears (2006) Social Isolation in America: Changes in core discussion networks over two decades, *American Sociological Review*, 71(3): 353–75.

Mewes, J. (2009) *Ungleiche Netzwerke – Vernetzte Ungleichheit: Persönliche Beziehungen im Kontext von Bildung und Status*, VS-Verlag, Wiesbaden.

Milgram, S. (1967) The small world problem, *Psychology Today*, 1(1): 61–7.

Mossong, J., N. Hens, M. Jit, P. Beutels, K. Auranen (2008) Social contacts and mixing patterns relevant to the spread of infectious diseases, *PLoS Med*, 5(3): 381–91.

Noelle, E. (1967) *Umfragen in der Massengesellschaft*. Einführung in die Methoden der Demoskopie, Rowohlt, Reinbek.

Otnes, P. (2009) A tie is not just a tie. From 'weak ties' to 'social capital', a critique, *Sosiologisk Arbok*, 3: 120–46.

Phillips, D.P. (1979) Suicide, motor vehicle fatalities, and the mass media: Evidence toward a theory of suggestion, *American Journal of Sociology*, 84(5): 1150–74.

Pool, I. and M. Kochen (1978) Contacts and Influence, *Social Networks*, 1: 5–51.

Portes, A. (1998) Social capital: Its origins and applications in modern sociology, *Annual Review of Sociology*, 24: 1–24.

Pustejovsky, J.E. and J.P. Spillane (2009) Question order effects in social network name generators, *Social Networks*, 31: 221–9.

Quarantelli, E.L. (1954) The nature and conditions of panic, *The American Journal of Sociology*, 60 (3): 267–75.

Roberts, S.G.B., R.I.M. Dunbar, T.V. Pollet and T. Kuppens (2009) Exploring variation in active network size: Constraints and ego characteristics, *Social Networks*, 31: 138–46.

Salentin, K. (1999) Die Stichprobenziehung bei Zuwandererbefragungen, *ZUMA-Nachrichten*, 23(45): 115–35.

Schnettler, S. (2009a) A structured overview of 50 years of small-world research, *Social Networks*, 31: 165–78.

Schnettler, S. (2009b) A small world on feet of clay? A comparison of empirical small world studies against best-practice criteria, *Social Networks*, 31: 179–89.

Schweizer, T., M. Schnegg and S. Berzborn (1998) Personal networks and social support in a multiethnic community of southern California, *Social Networks* 20: 1–21.

Scott, J. (2007) *Social Network Analysis*. A Handbook, Sage, Thousand Oaks.

Sen, S. and S. Bricka (2009) Data collection technologies – Past, present, and future, *Conference Paper for the 12th Meeting of the International Association for Travel Behaviour Research*, Jaipur.

Silvis, J., D. Niemeier and R. D'Souza (2006) Social networks and travel behaviour: Report from an integrated travel diary, *Conference Paper for the 11th International Conference on Travel Behaviour Research*, Kyoto.

Smieszek, T., M. Balmer, J. Hattendorf, K.W. Axhausen, J. Zinsstag and R.W. Scholz (2011) Reconstructing the 2003/2004 H3N2 influenza epidemic in Switzerland with a spatially explicit, individual-based model, *BMC Infectious Diseases*, 11: 115.

Sobel, J. (2002) Can we trust social capital?, *Journal of Economic Literature*, 40(1): 139–54.

Steglich, C., and T.A.B. Snijders (2010) Dynamic networks and behavior: Separating selection from influence, *Social Methodology*, 40(1): 329–93.

Trappmann, M., H.J. Hummell and W. Sodeur (2011) *Strukturanalyse sozialer Netzwerke. Konzepte, Modelle, Methoden*, Wiesbaden: VS Verlag.

Travers, J., and S. Milgram (1969) An experimental study of the small world problem, *Sociometry*, 32(4): 425–43.

Van den Berg, P.E.W., T.A. Arentze and H.J.P. Timmermans (2009) Size and composition of ego-centered social networks and their effect on travel distance and contact frequency, Transportation Research Record, 2135 1–9.

Wasserman, S. and K. Faust (2007) *Social Network Analysis: Methods and Applications*, Cambridge: Cambridge University Press.

Wellman, B. (1979) The community question: The intimate networks of East Yorkers, *American Journal of Sociology*, 84: 1201–131.

Wellman, B. and S. Wortley (1990) Different strokes from different folks: Community ties and social support, *American Journal of Sociology*, 96(3): 558–88.

Wolf, M. (2004) Egozentrierte Netzwerke: Erhebungsverfahren und Datenqualität, *Kölner Zeitschrift für Soziologie*, Sonderheft 44: 244–73.

Zhang, J., H.J.P. Timmermanns (2012) Activity-travel behaviour analysis for universal mobility design, *Transportmetrica*, 8: 149–56.

Chapter 5
Country Specific Characteristics Matter

Andreas Frei, Matthias Kowald, Pauline van den Berg and Juan A. Carrasco

Introduction

The analyses in the previous chapters of this book have demonstrated and analysed influences from personal networks on social interactions including travel (see Chapter 3) and a Swiss population-wide network (see Chapter 4). Besides the data collection efforts in Switzerland several similar efforts have been undertaken in different parts of the world with the motivation to investigate influences among socio-demographics, personal network characteristics, mobility biographical aspects, social network geographies and spatial patterns between social contacts. In addition to the data collection efforts in Switzerland (Chapter 2, also see Ohnmacht and Axhausen 2005; Chapter 3, also see Frei and Axhausen 2007; Chapter 4, also see Kowald and Axhausen 2012), there have been also collection efforts in Canada (Hogan et al. 2007; Carrasco et al. 2008), the Netherlands (van den Berg et al. 2009), and Chile (Carrasco and Cid-Aguayo 2012). These studies have highlighted similar observations and the relevance of different aspects of individuals' personal networks on their activity and travel behaviour.[1]

However, there is a need to disentangle the differences between these different data sets, since it is plausible that observed spatial patterns are influenced by the local, regional, or national context in which they are embedded or by the time of the study. In fact, our expectations are that the bulk of the contacts reported will be local/regional, as the cost structure for local and regional travel is comparable across the countries. In addition, given the relative affordability, one would expect the mean distances between social contacts to be higher in Switzerland, Canada and the Netherlands compared to Chile. Finally, the share of foreign nationals will influence the share of long distance and international contacts. As the cost for interactions with different means is not stable over time, one would also expect that the findings from these datasets are not easily transferable over the decade during which they were conducted.

Changes in accessibility have permitted travellers and residents to adjust the geography of their daily lives and their social networks (Hampton and Wellman

1 This chapter is partly based on the article 'Distance patterns of personal networks in four countries: A comparative study' (M. Kowald, P.E.W. van den Berg, A. Frei, J.A. Carrasco, T.A. Arentze, K.W. Axhausen, D. Mok and B. Wellman 2013).

2001). One of the strongest constraints is that the generalized costs of travel limit social contacts, and also the travel patterns. Changes in generalized costs of travel are therefore fundamental to changes in the social contacts as well as in travel patterns. Public investment in motorways and roads and private investment in vehicles have, for example, scaled Switzerland down from 1950 to 1990 for the road user by a factor of two and a factor of 1.5 for public transport. These costs did not fall further from the 1990 onwards, and they even increased in the last years through fuel costs and congestion (Axhausen 2006, see Figure 5.1).

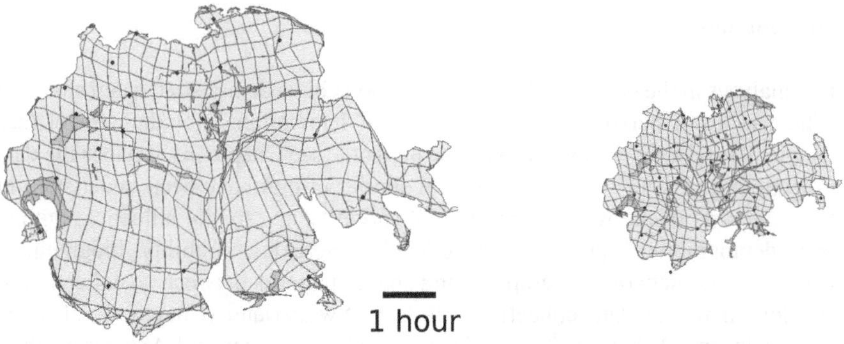

Figure 5.1 Road travel time: scaled maps of Switzerland (same scale for both years)

Note: Maps: Axhausen, Dolci, Fröhlich, Scherer and Carosio (2006); Scale: 1 hour.

Costs of car ownership and car operation have fallen from 1950 to 1990 even faster (Frei 2005 and Raff and Trajtenberg 1996) by a factor of 3.9, but from 1990 onwards only by 10% as illustrated in Figure 5.2. The collapse of long-distance telecommunication costs over the period of 1950 to 1990 has been similarly dramatic and scaled down the costs by a factor of 3.1, but from 1990 until 2001 it dropped even more, by a factor of 5.8 (see FCC 2001) and illustrated in Figure 5.2. This decrease of telecommunication costs is on-going and if we would include the voice-over-Internet-protocol (VOIP) telephony, the marginal monetary costs have become zero. So if we want to understand the effects of these dramatic changes in communication costs and take into account, that the generalized costs of travel did not change much since the nineties, the interactions of the different modes of contact in understanding the geographies of social networks are crucial. Therefore the change in social network distances over time might also be crucial. Our expectation is though that the most important form of contact to maintain a social network across countries and time are face-to-face contacts. But as their distances increase, other forms of contacts become more important than in the past. With the increasing availability of the internet over the last 10 years, we also expect that market-shares of contacts over the internet increased, especially at longer distances. In addition, given the relative affordability, one would expect the market-shares for face-to-face contacts at long distances to be higher in Switzerland, Canada and the Netherlands than in Chile. These dynamics in social networks are hard to observe. As there is no data available over a long period of time, data from different countries and data with very similar background but from different years might provide insight into how cost and accessibility change social network geography.

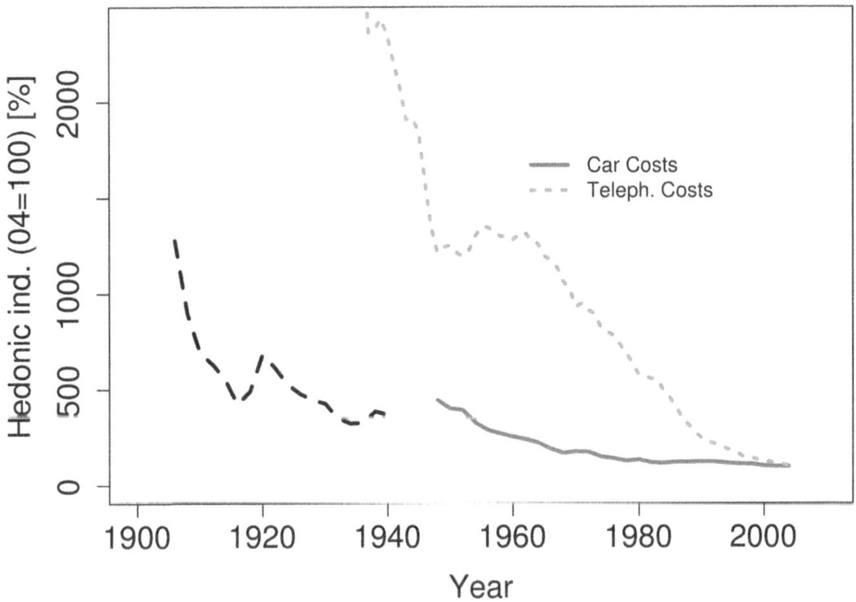

Figure 5.2 Quality adjusted 2004 purchase prices for private cars 1906–2004 and real costs of US interstate and international telephone calls 1930–2000

Note: Adapted from Frei 2005, Raff and Trajtenberg 1996 and FCC 2001.

With these motivations and expectations, the objective of this chapter is the comparative study of five datasets of social contacts from personal networks in different countries and/or different years.

The study is facilitated by the common personal network approach used, which focuses on specific individuals (egos) and their social contacts (alters). The section is structured as follows. In the next section, the five data collection efforts are described, followed by a descriptive analysis of the distances between ego and alter's home locations and then the contact modes used. Finally, the key conclusions are summarized.

Data Collection

In this chapter, five datasets from four different countries are used to compare patterns of personal social networks. The five data collections are described below, including the two datasets used in Chapters 3 and 4 of this book.

Toronto

In the East York area of Toronto, Canada, data were collected between May 2004 and April 2005 as part of the Connected Lives Study, a study about people's communication patterns. The study consisted of two stages: a random sample survey with 350 respondents from the East York area in Toronto, and interviews and observations of a sub-sample of them.

In these interviews, respondents were asked to name the persons who live outside their household, with whom they felt very close and somewhat close (for general information on the name generator technique see Chapters 3 and 4). Very close people consist of those persons with whom the respondent discusses important matters or regularly keeps in touch with, or are there for them if they need help. Somewhat close people were described as those persons who are more than just casual acquaintances, but not considered to be very close.

In total, 87 respondents completed the detailed questionnaire and named 1019 alters, i.e. an average degree of 11.7. Several attributes were collected for each alter, including home location, and frequency of contact by mode; more details can be found in Hogan et al. (2007) and Carrasco et al. (2008).

Zurich

Between December 2005 and December 2006 social network data were collected in Zurich, Switzerland (also see Chapter 3). Based on a random sample of the Zurich population, the participants were recruited on the telephone. The survey had two parts. First, a written questionnaire was filled in by the respondents containing socio-demographic and travel related questions. Second, a face-to-face interview was conducted to collect the social network.

Respondents were asked to name alters with whom they discuss important problems, with whom they stay in regular contact or whom they can ask for help. These questions cover the 'very close' or 'most important' contacts. The second name generator asked for persons with whom the respondents plan and spend leisure time. Both of these name generators targeted specifically social contacts influencing social activity and leisure behaviour.

In total, 307 respondents completed the questionnaire and named 3807 alters, i.e. with an average degree of 12.4. Similarly as in Toronto, several alter attributes were collected, among the exact home address location, and the contact frequency by mode. For more information, see Frei and Axhausen (2007) and Chapter 3 in this book.

Eindhoven

Between January and June 2008, social network data were collected in a number of neighbourhoods in the Dutch Eindhoven region. Eindhoven is a mid-sized city in the south of the Netherlands, with a population of 216,000.

The data were collected as part of a larger study, which consisted of a two-day social interaction diary (including a questionnaire on personal socio-demographic characteristics) and a follow-up questionnaire to capture the respondents' social network.

The participants of this social network study are a subset of the respondents who participated in the larger social interaction study. The social interaction study involved 747 respondents. A subsample of 116 respondents completed the social network questionnaire. In this study a paper and pencil questionnaire was used, in which respondents could self-report their social network members. The name generating questions that were used are similar to those used in the Connected Lives Study.

Respondents could record up to 25 very close and 40 somewhat close social network members. Among 116 respondents a total of 2,695 social network members were reported, resulting in an average degree of 23.28. For more information, see van den Berg et al. (2009).

Concepción

The study Communities in Concepción focused on the characteristics of social activity travel through the analysis of personal networks in different neighbourhoods of Concepción, Chile. The city is located 500 km south from Chile's capital, Santiago, and its Greater Area has a population of around one million people, being the second largest in the country. Data about personal networks were collected in four distinctive neighbourhoods as a way of capturing diverse income and accessibility to the CBD levels (see Chapter 6 for more details).

The data collection effort took place between August 2008 and April 2009. The data were collected in semi-guided interviews from 240 people (60 from each neighbourhood), who elicited a total of 5,053 personal networks members, which results in an average network size of 22.24. Respondents were chosen by a random and socio-demographic quota based procedure. The study used the same name generators as used in the Connected Lives Study, and included the network's spatial locations, frequency of interaction, social support exchange, and a two-day retrospective activity-travel survey. More details about this dataset can be found in Carrasco et al. (2013) and in Chapter 6.

Switzerland

Applying snowball sampling, data were collected on personal networks and a population-wide network structure between 2009 and 2011. A stratified random sample of the Canton Zurich population was used to recruit 40 initial respondents. Two name generators aimed for leisure and emotionally important contacts, asking respondents to mention people with whom they make plans to spend free time and for people that are used to discuss important problems. The questionnaire provided space for 40 names and encouraged respondents to write down additional names on an extra sheet of paper if needed (see Chapter 4).

Employing snowball sampling means to use all contacts mentioned in response to the name generator as the basis for further recruitments. In other words, these persons were asked to fill out the paper-based questionnaire as well. Repeating this process on four iteration levels resulted in a sample of connected personal networks containing information on 793 egos and 15,593 alters. Although recruitment efforts were taken to the international level, respondents and their social contacts are highly clustered in Switzerland. The data include important information on a population-wide leisure network structure (for more information see Kowald and Axhausen 2012 and Chapter 4 in this book).

Dataset Comparison

A challenge of the analysis of the datasets was identifying the common variables available in all five datasets and defining common categories accordingly. In this process, there were some survey specific characteristics that needed special treatment. First, the Switzerland data have a large number of missing values in terms of distances between respondents' and their social contacts' home locations, which is a direct result of its snowball sampling strategy. Respondents' that did not want their social contacts to become part of the sample refused mentioning their postal addresses. Second, distances were surveyed with an ordinal scale in Eindhoven. Only distances above 200 km were recorded on a metric scale. However, in order to compare distances on a metrical scale, all values within the distance classes were replaced by the average within the class boundaries. Third, distances were geo-coded in Toronto, Zurich, Switzerland, and Concepción using egos' and alters' addresses or in cases where the address was not available (7% in Zurich) the next best guess from the respondent (street corner, closest landmark etc.). Contrarily, the Eindhoven dataset includes distances as reported by respondents. This limits the comparison of the spatial structure of the ego's social network to straight-line distances for each ego-alter pair. Alternatives that take the clustering effect of alters into account (see Schönfelder 2006 for a comparison of different metrics; see Carrasco et al. 2006 and Frei and Axhausen 2007 for the analysis of ego's network's spatial representation with confidence ellipses) are not available in this comparison, but have been shown to correlate strongly with crow-fly distance (see Frei and Axhausen 2007). The available data does also not account for cultural clustering or population density. For the contact modes it has to be mentioned that texting (SMS) frequencies were only collected in the case of Zurich and Switzerland and Eindhoven. In Eindhoven and Concepción the mode contact frequencies were collected as categories, whereas in Switzerland, Zurich and Toronto they were collected as continuous numbers. The coarsest category of Concepción was chosen for comparison, where the frequencies were collected as once a year or less, between once a month and once a year, between once a week and once a month and weekly and more frequent.

For the purposes of the analysis, observations with missing values were excluded from the datasets. A caveat of this approach is that the remaining observations may not be entirely representative within each dataset.

The descriptive analysis provides a comparative overview of the datasets, highlighting the dissimilar target populations in the four countries, as well as the difference on the name generators and information collected at the several units of analyses, all of which may have some impact on the results.

Table 5.1 Location specific statistics (2006)

	Toronto	Zurich	Eindhoven	Concepción	Switzerland
Population [in 1000]	7,866	372	213	292	7,866
Population density / km^2	3,972	4,049	2,407	1,318	188
Wage level (Base: New York) Gross [kUS$]	74.2	115.1	77.0	21.2	115.1
Wage level (Base: New York) Net [kUS$]	80.4	124.2	72.7	24.3	124.2
Bus/Tram/Metro (Network ticket for a 10 km trip) [US$]	2.4	2.7	2.6	0.7	2.7
Taxi (per 5km) [US$]	8.2	21.2	17.2	0.7	21.2
Train (200 km single ticket)	45.4	44.8	31.2	11.7	44.8
Average [US$]					
Average cost of fuel per litre [US$]	0.89	1.22	1.72	1.06	1.22
Average Mid-Price Car [US$]	19,933	22,240	21,140	11,416	22,240
Tax on Car [US$/ Year]	64	255	289	210	255

Source: UBS, Wealth Management Research (2006).

Table 5.1 provides an overview of each survey. All datasets were surveyed in urban areas, except the Switzerland study, which collected information at the national level. National statistics from the three developed countries show small differences in wage level and transportation costs, especially if their ratio is compared. The Chilean case of Concepción is different, as the wage level is much lower, and the lower costs for individual transport do not cancel out the wage difference. Finally, note that the Swiss national numbers shown in Table 5.1 may understate the incomes in Zurich.

Comparison between Respondents' Socio-Demographics and Network Characteristics

Table 5.2 provides a comparison between egos' characteristics. The different datasets are quite similar and show only few major differences. The Concepción survey population is younger than those of the other surveys, which results from a high share of participants in the youngest age class (<30 years). This is related to the phenomenon of the demographic change, describing decreasing birth rates and an older overall population in more developed western countries, which has not been observed in Chile. In terms of household structure, the Zurich data show a nearly perfect split between respondents living with or without partner whilst there is a pro-couple dominance in all other data sets. Another difference concerns the presence of children in the respondents' households. Households with young children represent a third in Eindhoven and Switzerland, half of the sample in Toronto, and two thirds in Concepción. In terms of the educational level, the four data sets from the OECD countries show low shares for primary education, and – with the exception of Zurich – higher shares for educational levels with academic degrees. The income distribution – defined in terms of low, medium and high income categories within each country – follows different patterns in each data sets. The survey population from Concepción has more people in the lowest income class, whilst the Switzerland study is dominated by respondents in the highest income class. Most households in the western world own a car, which is also true for Concepción, even though the dominance of car ownership in this latter case is less strong. No matter where respondents' households are located, most of them have Internet access. In addition, most respondents have a mobile phone, with the exception of the Toronto data set and Zurich, which are the oldest datasets used in this comparison. In terms of years living in the current location, the Switzerland data show a higher average number than the other data sets.

Table 5.2 Socio-demographic characteristics of the respondents

	Toronto	Zurich	Eindhoven	Concepción	Switzerland
Number of respondents	84	265	106	241	426
	Sample (%)	*Sample (%)*	*Sample (%)*	*Sample (%)*	*Sample (%)*
Male	39.8	42.3	31.1	39.8	41.1
Young (<30)	9.8	19.6	7.5	24.1	5.9
Middle (30–60)	69.5	44.2	59.4	58.5	73.5
Old (>60)	20.7	36.2	33.0	17.4	20.6
Living with partner	61.9	48.3	72.6	58.1	80.3
Child(ren) under 18	46.4	–	34.0	61.8	36.6
Primary education	18.3	7.9	17.0	44.8	1.6
Secondary education	28.0	70.2	34.0	24.5	48.6
Higher education	53.7	21.9	49.0	30.7	49.8
Low HH–income	29.4	24.5	36.8	43.8	10.1
Medium HH–income	41.2	47.9	25.5	25.1	39.2
High HH–income	29.4	27.5	37.7	31.1	50.7
1 or more cars	–	63.4	83.0	56.4	89.0
Season ticket	–	38.9	42.5	–	82.9
Internet access	79.8	67.9	90.6	63.9	97.9
Mobile phone access	42.3	65.7	94.3	86.3	95.8
	Mean	*Mean*	*Mean*	*Mean*	*Mean*
Age	50.3	50.7	51.6	42.8	50.1
Work hours	22.5	–	14.6	21.8	–
Years in current location	12.8	–	13.3	16.4	25.4

Source: Kowald et al. (2013).

Table 5.3 describes the characteristics of the social contacts in each data set. In terms of personal network sizes, the mean value is very similar among all data sets, with the exception of Zurich, where the value is smaller. This similarity may be in part due to the same kind of name generators used by all data sets in terms of emotional closeness, although the Zurich and Switzerland datasets also added leisure contacts as another criterion. Another hypothesis about this similarity could be that there may be a 'cognitive' threshold that causes the similar network sizes; however, the results from Zurich complicate this idea. In addition, the difference between the Zurich and Switzerland datasets could result from differences in the name generator asking for leisure contacts is more detail in the latter survey. This hypothesis supports the idea that name generators are very sensitive to (even small) changes. The previous mixed evidence about network sizes suggests the need of more research to better understand these issues.

Table 5.3 Characteristics of personal networks

	Toronto	**Zurich**	**Eindhoven**	**Concepción**	**Switzerland**
Number of alters	1,019	3,156	2,452	5,038	7,293
Average network size	23.8	11.9	23.9	20.9	21.6
	Sample (%)	*Sample (%)*	*Sample (%)*	*Sample (%)*	*Sample (%)*
Male	42.1	–	41.4	45.9	43.1
Young (<30)	11.8	–	13.0	32.2	9.2
Middle (30–60)	65.3	–	53.7	52.9	67.2
Old (>60)	22.9	–	33.3	14.9	23.6
Immediate family	25.4	18.3	18.8	20.2	13.4
Extended family	11.0	12.7	25.1	23.6	11.1
Friend or other	63.6	69.0	56.1	56.2	75.5
Very close	54.0	52.4	43.0	51.3	28.9
Somewhat close	46.0	47.6	57.0	48.7	71.1
Known <1 year	–	0.6	1.4	14.6	2.0
1–10 years	–	36.7	37.4	25.8	28.5
>10 years	–	62.7	61.2	59.6	69.5

Source: Kowald et al. (2013).

Regarding the alters' age distribution, it shows similar patterns to egos' age. Whilst a comparison between Toronto, Eindhoven, and Switzerland fits well, the share of social contacts in the youngest age class for the Concepción study is larger. Since the Zurich survey did not include questions on alters' sex and age, this dataset cannot be used to calculate corresponding homophily values.

In terms of the ego-alter relationships, Switzerland and Zurich include a high share of friends and only a low share of family members. In the case of Eindhoven and Concepción, the friendship category is dominant as well, but the distribution is more balanced with respect to family members. Toronto situates somewhere between these two patterns. These differences can be, at least partly, due to the different name generators employed.

Regarding emotional closeness, the share of very close contacts is very similar across the datasets, with the exception of Switzerland, which has a much lower share, possibly due to its stronger orientation towards leisure contacts.

Finally, in terms of the relationship duration, the data sets follow different patterns. A particular noteworthy case is Concepción, where the relationships are much younger, which could be explained in part by the age distribution in this dataset.

Spatial Distribution of Social Contacts Comparison

Figure 5.3 and Tables 5.4 and 5.5 show the general picture of the spatial distribution of the different personal networks. The geo-referenced values of the respondents' homes and their social contacts were used to calculate great circle distances, employing an equidistant cylindrical projection to account for the shape of the Earth. Eindhoven is excluded in this analysis since the distances were reported in classes only. The distribution of the great circle distances (see Figure 5.3c) between the respondents' homes and their social contacts has three elements. Around two-thirds of the alters live locally within 30 km (roughly 30 minutes to 1 hour travel time by car) (Figure 5.3a). The remaining distances are roughly equally divided into regional and national relationships (within 30–100 km) (Figure 5.3a) and longer distance and international relationships (>100 km) (Figure 5.3b). The 100 km threshold was chosen to mark the difference of long-distance travel and everyday travel, as this is a common cut-off point in transportation planning See for example Axhausen, Madre, Polak and Toint 2002). Overall, the respondents mix their daily life local/regional contacts with a multitude of non-local and often long distance contacts. In fact, there is also a noticeable high share of intercontinental links, especially in the case of Toronto, possibly due to immigrants maintaining their links to their birthplace.

In the spatial range of everyday life (less than 100 km), the overall differences in the distance distributions between the datasets are rather small. For longer distances, there are no clear visible trends that can be revealed by the aggregated descriptive analysis presented here, except regarding the peaks in the longer

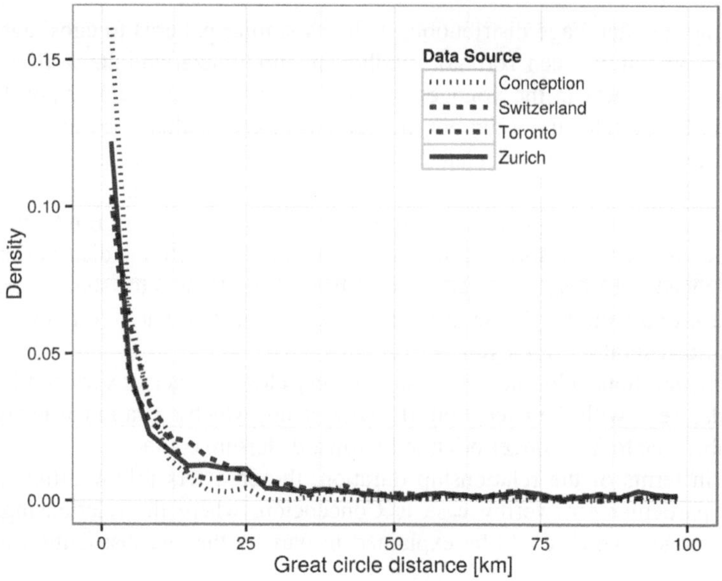
Data Source
Conception
Switzerland
Toronto
Zurich
Density
0.15
0.10
0.05
0.00
0
25
50
75
100
Great circle distance [km]

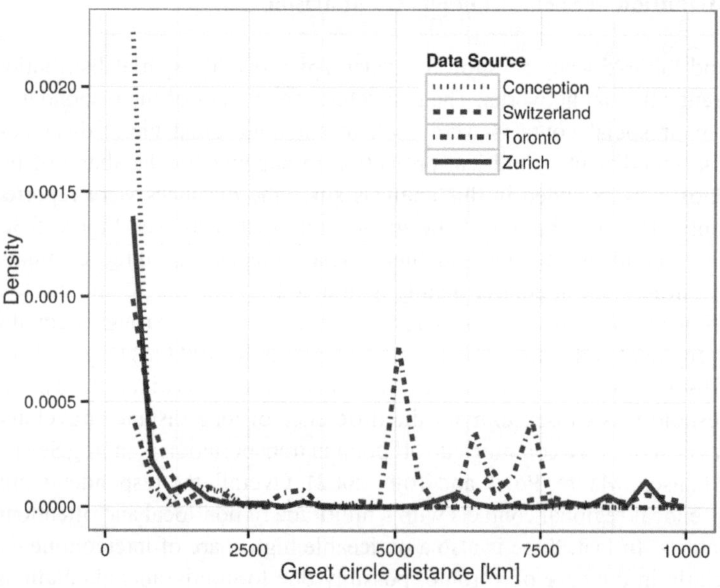
Data Source
Conception
Switzerland
Toronto
Zurich
Density
0.0020
0.0015
0.0010
0.0005
0.0000
0
2500
5000
7500
10000
Great circle distance [km]

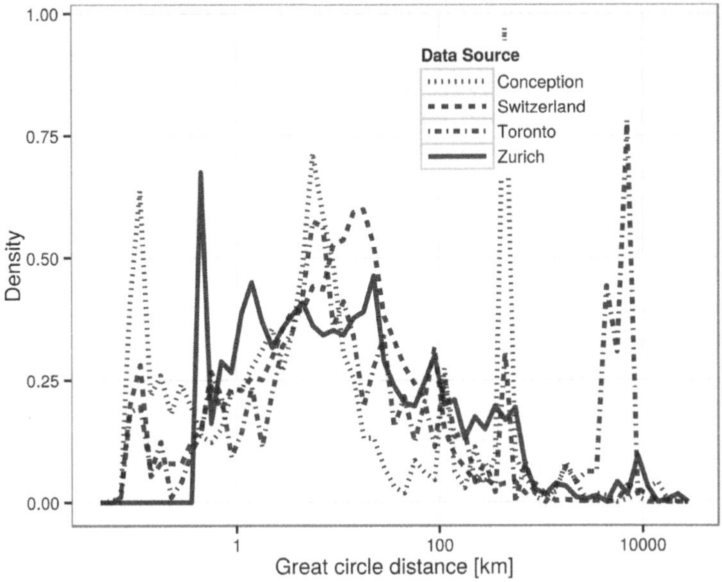

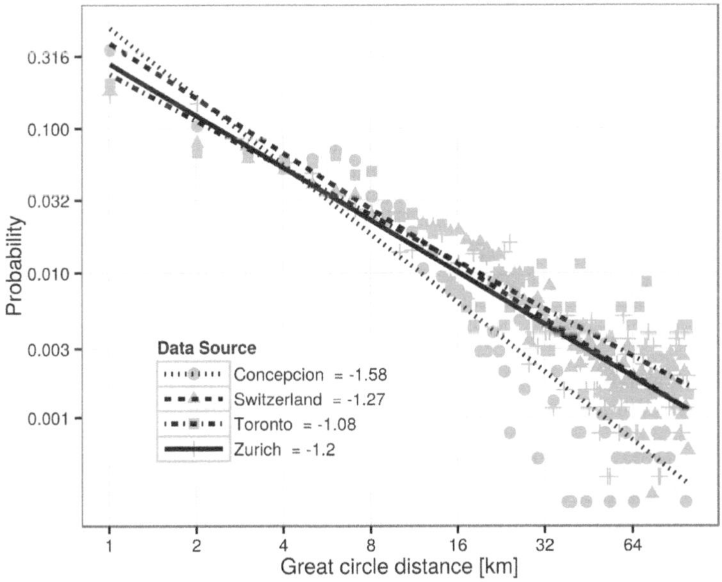

Figure 5.3 Distance distribution between egos and alters a) Up until 100 km b) Over 100 km c) Over the whole distance spectrum d) Relationship probability

distances for Toronto. Figure 5.1d presents the tie distance distribution within 100 km in more detail, showing the tie distance cut in one km bins and their shares on a log-log scale. The tie length distribution follows a power law for all four datasets, defined by $p_{tie} \sim distance^{\beta}$, where β varies between −1.08 (Toronto) and −1.58 (Concepción). The order of the datasets is as expected, given their background. In fact, transportation costs compared to the wage levels are similar in Zurich, Switzerland and Toronto; also this latter dataset includes immigrants with a higher share of longer and shorter ties. The decay in the tie probability with distance is larger in Concepción, which could be explained in part due to relatively higher transportation costs.

Table 5.4 gives the power law estimates and the tie distance comparison between the different datasets. As seen before, Concepción has the strongest decay of social ties with distance followed by Switzerland, Zurich and Toronto. The difference in the relationship between the number of ties and distance is statistically significant between the datasets, except for the tie distance relationships between the datasets from Zurich and Switzerland, which are very similar.

It is also interesting to note that in all data sets there are social contacts living at distances larger than 100 km, 1,000 km, or even 10,000 km. A local peak can clearly be identified at around 8,000 to 12,000 km, resulting from the distance between America and Europe (see Figure 5.3c).

Table 5.4 Estimates of the power law parameter by study area for ties under 100 km distance

Distance power law β	Est.	Std.E.	P(> |t|)	Est. diff. from Concepción	P (> |t|)	Est. diff. from Switzerland	P (> |t|)
Concepción	−1.58	0.07	0.000	–	–	–	–
Switzerland	−1.27	0.06	0.000	0.31	0.001	–	–
Zurich	−1.20	0.06	0.000	0.38	0.000	0.07	0.387
Toronto	−1.08	0.07	0.000	0.50	0.000	0.19	0.032

Source: Kowald et al. (2013).

The data from Eindhoven were excluded in Figure 5.3 considering that distances were recorded in discrete ranges as already mentioned before. However, Table 5.5 presents the comparison of the Eindhoven data with the other four datasets.

Table 5.5 Distance distribution of social contacts by study area

	Toronto	Zurich	Eindhoven	Concepción	Switzerland
Distance < 2km	19.3	26.0	17.9	36.1	24.6
Distance 2–30km	42.9	43.2	47.6	41.3	54.1
Distance 30–100km	9.4	12.4	17.3	2.4	14.5
Distance > 100km	28.3	18.4	17.2	20.2	6.8
Distance mean	1,036.0	286.6	152.9	222.9	106.3
Distance median	11.2	9.0	10.0	4.9	8.9

Source: Kowald et al. (2013).

Contact Frequency and Contact Mean Comparison

Table 5.6 shows the characteristics of contact frequencies by level of emotional closeness divided by the three contact means: face-to-face, telephone and internet. For the face-to-face contact frequencies it can be observed that there is little difference between very close relationships to somewhat close ones. In the Zurich and Switzerland dataset, the very close relationships tend to be maintained more frequently compared to the somewhat close relationships, whereas this is the opposite case for Eindhoven, Toronto, and Concepción. In general, Concepción shows much higher frequencies compared to the other countries. This could be a cultural difference, but could also be due to the fact, that tie distances are shorter in the case of Concepción, and therefore Concepción has lower average costs for face-to-face interactions. Face-to-face interaction frequencies are in general lowest for Zurich and Switzerland, which could represent again cultural differences. Similar to face-to-face contacts, the differences between emotional close and somewhat close relationships are small for telephone contacts and internet contacts. Again, Switzerland and Zurich show increased communication frequencies for emotional close contacts, whereas the opposite is true for the other datasets. Overall, the shares of rare contact frequencies are much higher for telephone and internet contacts compared to face-to-face contacts, especially in the case of Concepción. This could be represented by the accessibility and the overall higher relative costs of electronic contact forms in Concepción. Overall there is little difference in contact frequency between the different contact means by type of relationship.

Table 5.6 Contact frequency and emotional closeness by study area

Emotional closeness	Dataset	Once a year or less (%)	Between monthly and yearly (%)	Between weekly and monthly (%)	More than weekly (%)
Face-to-face contacts					
somewhat close	Eindhoven	0.85	31.69	33.11	34.35
	Zurich	13.97	45.11	22.75	18.16
	Switzerland	7.42	58.80	26.74	7.04
	Concepcion	0.00	19.60	24.19	56.21
	Toronto	9.56	52.73	2.19	35.52
very close	Eindhoven	1.65	44.42	29.33	24.61
	Zurich	8.36	43.97	27.92	19.75
	Switzerland	2.63	44.89	30.96	21.52
	Concepcion	0.00	22.42	27.50	50.09
	Toronto	8.41	59.81	1.87	29.91
Telephone contacts					
somewhat close	Eindhoven	6.64	24.00	38.99	30.36
	Zurich	24.15	39.06	25.88	10.91
	Switzerland	27.45	54.45	14.27	3.84
	Concepcion	15.81	19.13	21.98	43.08
	Toronto	3.28	42.35	1.64	52.73
very close	Eindhoven	13.52	57.58	23.53	5.36
	Zurich	20.53	31.01	33.56	14.90
	Switzerland	8.82	37.62	37.31	16.25
	Concepcion	30.58	23.90	18.82	26.70
	Toronto	22.74	55.76	1.87	19.63
Internet contacts					
somewhat close	Eindhoven	40.80	26.94	25.62	6.64
	Zurich	68.33	19.03	8.85	3.79
	Switzerland	53.17	34.55	9.40	2.88
	Concepcion	72.45	8.74	5.06	13.75
	Toronto	50.00	23.50	2.73	23.77
very close	Eindhoven	55.58	26.47	14.59	3.36
	Zurich	66.75	17.32	11.14	4.78
	Switzerland	45.51	29.72	19.04	5.73
	Concepcion	70.56	10.04	6.39	13.01
	Toronto	55.14	28.35	2.18	14.33

Figure 5.4 shows the shares of the modes as a function of the distance between the respondents and their contacts in distance deciles. It can be seen that the distance decay of contact frequency varies considerably by mode, which will lead to rather different conclusions when the modal shares are analysed compared to the absolute numbers presented in the previous section. The share of the face-to-face visits decreases quickly. Email gains share fastest with increasing distance and stays constant in terms of frequency over the distance range. Phone calls gain share as well for all datasets besides Switzerland. Phone calls even overtake face-to-face contacts at around 100 km distance in terms of frequency for all datasets other than Switzerland. The decay in market share for face-to-face contacts with increasing distance is similar for all datasets but for Eindhoven, where the decay is a much stronger.

Telephone and internet market-shares grow fastest in the Eindhoven case, whereas Switzerland's internet contact shares grow similarly fast but with much lower share of phone contacts. The strong growth of internet contact market share in the case of Switzerland seems to be a substitute for the low phone market shares. This is expected as the Switzerland dataset is the newest, where internet accessibility and availability is higher compared to the dataset of Zurich and Toronto – collected seven years earlier. The growth of internet contact market share with distance is smaller in Concepción, which could be explained in part due the lower rate of internet accessibility and the higher relative internet cost over wage ratio, compared to the other study areas.

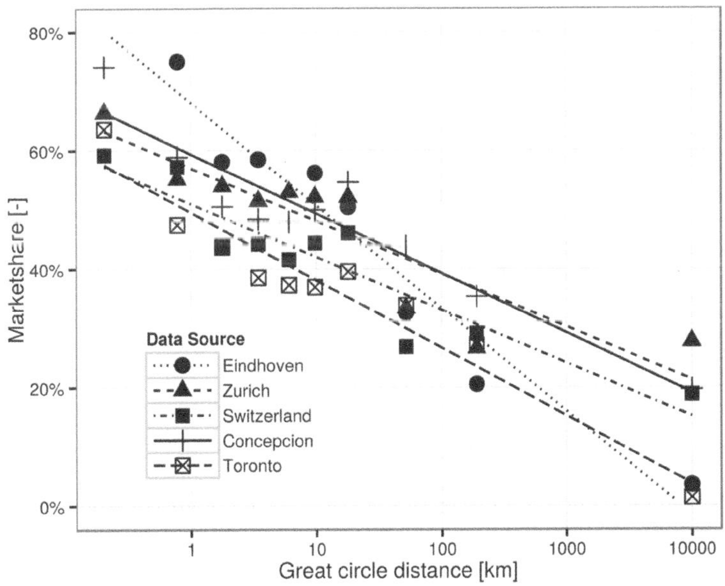

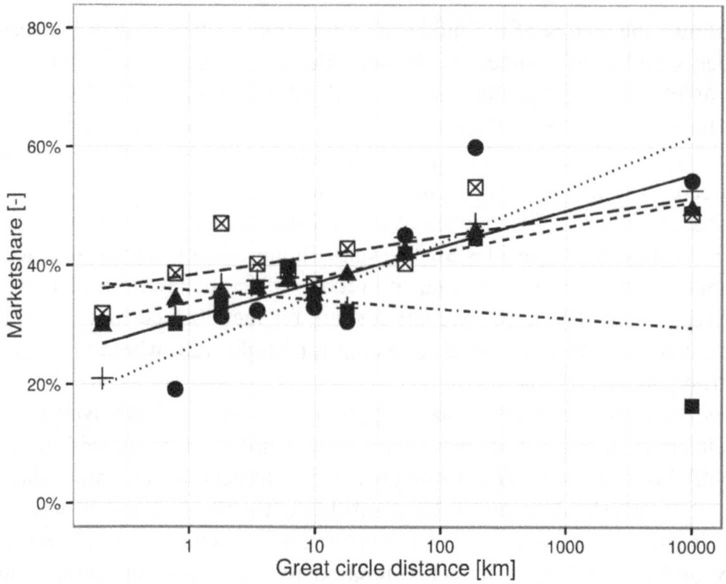

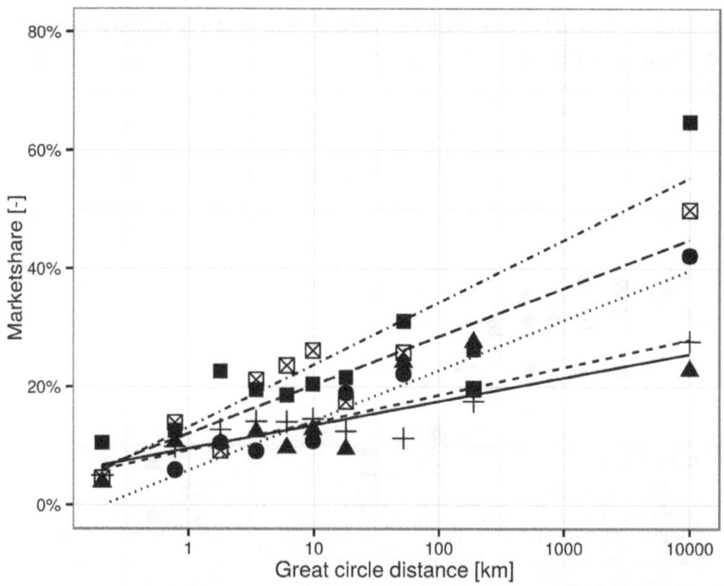

Figure 5.4 Share of the contact mode by log distance in deciles between the ego's and the alters' homes a) Face-to-face market-shares b) Telephone market-shares c) Internet communication market-shares

Conclusions

The interest in understanding the role of social networks in travel behaviour has motivated dedicated data collections in several countries that have used techniques from sociology and other related fields to elicit the respondents' personal networks. In this context, a key research question is related to the spatial distribution of individuals' social contacts, which constitutes a relevant portion of their social activity space, and is important to understand social activity-travel. Although previous chapters have analysed empirical data on this issue in Switzerland (see Chapters 2, 3, and 4), the aim of this chapter was performing a comparative exercise among five datasets from four different countries: Canada (Toronto), Chile (Concepción), Switzerland (Zürich and the whole country), and the Netherlands (Eindhoven). This comparative effort is facilitated by the similar approach of the five data collection efforts, both in terms of the social network eliciting techniques as well as the key explanatory variables explored.

The descriptive analysis highlighted the similarities and differences on the data collected from the different countries, in particular regarding the tie distance distribution. In fact, although all data sets follow a power law distribution, the strongest decay was found in Concepción, the slowest in Toronto, while the European datasets are in between these two cities. These differences highlight the relevance of contextual aspects such as the ratio between wage and transport costs, the availability of mobility tools – such as car ownership and internet access – and the influence of immigration.

Overall, the results suggest that the availability of transport and communication relative to income plays a role on the spatial distribution of contacts and how those contacts are maintained. Despite the context of the country, distance plays a very strong role, not only for face-to-face contacts, but in general. Substitution patterns for different contact modes become important as soon as distances surpass daily mobility patterns with distances over 100 km. However, besides these general regularities, the relevance and magnitude of the specific components strongly depends on their city or national context into which these networks are embedded.

References

Axhausen, K.W. (2006) Moving through nets: An introduction, in Moving Through Nets: The Physical and Social Dimensions of Travel, ed. K. W. Axhausen, 1–7, Oxford: Elsevier.

Axhausen, K.W., J.-L- Madre, J.W. Polak and Ph.L. Toint (eds) (2002) *Capturing Long Distance Travel*, Baldock, UK: Research Science Press.

Carrasco, J.A., B. Hogan, B. Wellman and E.J. Miller (2008) Collecting social network data to study social activity-travel behaviour: An egocentric approach, *Environment and Planning B*, 35(6): 961–80.

Carrasco, J.A. and B. Cid-Aguayo (2012) Network capital, social networks, and travel: An empirical illustration from Concepción, Chile, *Environment and Planning A*, 44(5): 1066–84.

Carrasco, J.A., C. Bustos, and B. Cid-Aguayo (2013) Affective personal networks versus daily contacts: Analyzing different name generators in a social activity-travel behaviour context, in Transport Survey Methods: Best Practice for Decision Making, ed. J. Zmud, M. Lee-Gosselin, M. Munizaga and J.A. Carrasco, 409-426, London: Emerald Press.

FCC (2001) Long distance telecommunications industry, Technical report, FCC, Washington, DC.

Frei, A. (2005) Was hatte man 1960 fur einen Sharan bezahlt?, Msc thesis, Department Bau, Umwelt und Geomatik, ETH Zurich, Zurich.

Frei, A. and K.W. Axhausen (2007) Size and structure of social network geographies, *Arbeitsberichte Verkehrs- und Raumplanung* 439, IVT, ETH Zürich, Zürich.

Hampton, K.N. and B. Wellman (2001) Long distance community in the network society: Contact and support beyond netville, *American Behavioral Scientist*, 45(3): 477–96.

Hogan, B., J.A. Carrasco and B. Wellman (2007) Visualizing Personal Networks: Working with Participant-aided Sociograms, *Field Methods* 19(2): 116–44.

Kowald, M. and K.W. Axhausen (2012) Focussing on Connected Personal Networks: Selected Results from a Snowball Sample, *Environment and Planning A*, 44(5): 1085–100.

Kowald, M., P.E.W. van den Berg, A. Frei, J.A. Carrasco, T.A. Arentze, K.W. Axhausen, D. Mok and B. Wellman (2013) Distance patterns of personal networks in four countries: A comparative study, *Journal of Transport Geography*, 31: 236–48.

Ohnmacht, T. and K.W. Axhausen (2005) Entwicklung des Forschungsdesign und der Erhebungsinstrumente für das Projekt Mobilitätsbiographien, Mobilitätswerkzeuge und soziale Netze, *Arbeitsberichte Verkehrs- und Raumplanung*, 298, IVT, ETH Zürich, Zürich.

Raff, D. and M. Trajtenberg (1996) Quality-adjusted prices for the American automobile industry: 1906–1940, Working Paper.

Schonfelder, S. (2006) Urban rhythms: Modelling the rhythms of individual travel behaviour, PhD thesis, ETH Zurich, Zurich.

Van den Berg, P.E.W., T.A. Arentze and H.J.P. Timmermans (2009) Size and composition of ego-centered social networks and their effect on travel distance and contact frequency. *Transportation Research Record*: Journal of the Transportation Research Board, 2135, 1–9.

Chapter 6
Personal Characteristics, Social Contacts and Transport Systems

Juan A. Carrasco

Introduction

The purpose of this section is to provide an international overview which complements the Swiss case study, by presenting a description of the characteristics of personal networks on two very different cities: Toronto, Canada, and Concepción, Chile. As we will see, patterns of spatiality and frequency of interaction on the personal networks on these cities are shaped in part by the specific cultural and economic contexts where they belong. These differences are further shaped by the specific transport system where they are embedded, particularly the travel demand and supply that support these social relations.

This perspective contrasts with traditional approaches, in which transport systems are studied through an economic-based lens, which concentrates almost uniquely on the user's socio-demographic attributes as a way of explaining travel behaviour. Although trip purposes such as commuting and shopping can be explained in part by segmenting the travellers, socio-demographics are not enough to understand other more complex but also important trip purposes, such as those mainly motivated for social contacts. A key way of broadening that insufficient theoretical scope is by characterizing the personal networks of travellers as a way of understanding the role of time (frequency of interaction) and space (distance between social contacts) on socializing behaviour.

This chapter describes the temporal and spatial patterns of personal networks in Toronto, Canada, and Concepción, Chile, as a way of eliciting hypotheses about the characteristics of social travel for different population groups in these two countries, providing an international overview, which will complement the Swiss case. After presenting a brief overview of the context in Toronto and Concepción, the rest of this section will present and discuss key characteristics on both cities of their personal networks, spatiality and the patterns frequencies of interaction.

Data Collection and Sampling

Carrasco et al. (2008a) report the method and key characteristics of the data collected in Toronto, Canada, while Carrasco et al. (2013) do the same for the data collected in Concepción, Chile. Toronto is the largest city in Canada, with 5.5 million inhabitants at the time of the data collection (2004); sampling focused on the neighbourhood of East York, which is located in the centre-east area of the city, with good accessibility to the subway and bus system, and relatively close to the Central Business District (CBD) area. The area is very diverse in terms of income and ethnic groups, this latter and important aspect in a city so diverse on ethnical backgrounds. Concepción is the second largest city in Chile, with one million inhabitants at the time of the data collection (year 2008); sampling focused on four specific neighbourhoods, which tried to represent the combinations of low and high income, on the one hand, and being close or far (with respect to walk) to Concepción's CBD.

Social Environment: Network Characteristics

Both data collection efforts elicit the emotional closest subset of the individual's social contacts through name generators, as discussed in previous chapters (see Chapters 3 and 4) and by Hogan et al. (2007) and Carrasco et al. (2008a). In addition, the respondents from these two cities were asked about the relationships among their social contacts, capturing the node-tie structure of their personal networks. In the case of Toronto, these ties were gathered for a subsample of the alters elicited, whereas in Concepción, these ties were gathered for all the alters that were named.

Using this node-edge configuration, several indicators of each respondent's social environment can be measured:

1. Size: number of alters elicited.
2. Density: proportion of existing ties with respect to the theoretical maximum.
3. Components: indicator of the number of subgroups existing in the network.
4. Isolates: alters without any other connection than the ego.
5. Degree of centrality: indicator of the average number of ties between an alter and the others (Freeman 1979).
6. Degree of betweenness: indicator of the paths that link any pair of alters through a third social contact (Freeman 1979).

Tables 6.1 and 6.2 present the network structure characteristics for Toronto and Concepción, respectively. Interestingly, the mean network sizes are very similar on both cities, with a slightly higher number in the case of Toronto. On both cities, there is a high variability on these values, with ranges of 3–66 and 5–72 alters, respectively, which can also be seen on their high standard deviations.

The similarity on network sizes contrasts with the differences on personal network density, where respondents from Concepción tend to have more alters who know to each other. Complementary, Torontonians have more fragmented networks in average, which may involve a richer set of social foci (Feld 1981) and

Table 6.1 Network structure characteristics – Toronto

	Network size	Sample size (only those from the mini survey)	Density (includes isolates)	Density (without isolates)	Number of components (includes isolates)	Number of components (without isolates)	Number of isolates
Mean	23.76	12.13	0.17	0.30	8.69	4.25	4.42
Mode	13	14	0.11	1.00	8	2	1
Std. Deviation	14.48	3.17	0.17	0.28	6.19	3.24	4.40
Minimum	3	3	0.02	0.05	1	1	0
Maximum	66	15	1.00	1.00	33	20	23
Percentile							
25	13	10.25	0.08	0.11	4.00	2.00	1.00
50	20	13.00	0.11	0.18	8.00	3.50	3.00
75	31	14.00	0.18	0.37	10.00	5.75	6.00

Table 6.2 Network structure characteristics – Concepción

	Network size	Density	Number of components	Number of isolates
Mean	20.93	0.24	5.60	2.52
Mode	15[a]	0.11[a]	2	0
Std. Deviation	10.17	0.19	4.61	3.03
Minimum	5	0.01	1	0
Maximum	72	1.00	35	19
Percentile				
25	14	0.11	2.0	0.0
50	19	0.18	5.0	2.0
75	26	0.31	7.0	3.5

Note: [a] Multiple modes exist. The smallest value is shown.

social resources, aspect that is reinforced by the higher number of isolate alters that, in average, respondent from Toronto present of their networks.

These results not only have potential consequences from a socializing viewpoint, but also can imply some aspects from a travel behaviour perspective. In fact, if every ego-alter tie represents a potential travel for social purposes, less dense networks and more subgroups on personal networks may involve potentially more social trips. In fact, having more subgroups requires having more different social activities to interact with others in time and space, since face to face contact remains as the key mode to maintain personal network contact. In other words, these personal network characteristics constitute structural conditions needed to understand the individual's travel demand for social activities.

Yet social space does not exist in isolation, and these structural conditions are mediated by people's behaviour in time and space. For that reason, we need to understand both the frequency of interaction as well as the spatial patterns (distances) of these ego–alter interactions, as a way of having a sense of social activity trip production and distribution. Next subsections will examine these aspects.

Distance and Personal Networks: Spatial Characteristics

Table 6.3 shows the statistics of the egos-alter home distances. The mean ego-alter home distance in Toronto is five times higher than Concepción, with 1036 and kilometres, which could be a reflection, in part, the high migration levels in the Canadian city. However, on both cities, the distance distribution function is highly skewed towards small values; in fact, half of the alters live at distances equal or closer to 11.2 kilometres in Toronto, and 4.5 kilometres in Concepción, showing that most of the personal networks are located within the urban limits. The differences between the two cities are sharper at the percentile 75, where some Torontonian egos have alters who live at 160 kilometres; in Concepción, this figure is only 16 kilometres.

In both cities, there is a high variability on distance, as can be seen by the high standard deviation, which reflects the spatial diversity of social contacts, and consequent travel behaviour. In fact, if we only concentrate on home-based social activity-travel, from a spatial viewpoint, there are high levels of local-based trips (at the neighbourhood and city scale), but also other potential trips that can be longer than the average commuting on these cities.

Complementing the previous results, Figures 6.1 and 6.2 present the frequency distribution function for Toronto and Concepción, respectively. Interestingly, for distances shorter than 100 kilometres, both cities present the same distance decayed, logarithmic, shaped function. Yet, in the case of Concepción, most of alters fall within the 30 km threshold, whereas there is proportionally a higher number of people even at 80 kilometers of distance in Toronto. Although a first potential source of explanation of these results is related to the higher area covered by the Greater Toronto Area, this difference can also reflect the existing people's mobility possibilities in order to maintain their personal networks. In fact, as discussed elsewhere (Carrasco 2011; Carrasco and Cid-Aguayo 2012), the availability of mobility tools (information and communication technologies, cars, transit) as well as how monetary and cognitive resources, become an difference on people's ability to maintaining social contacts (Lyons and Kenyon 2003).

Table 6.3 Distances between ego and alters per network – Toronto and Concepción

Ego-alter distance (km)	Toronto	Concepción
Mean	1,036	210
Mode	0	0
Std. Deviation	2,211	1,173
Minimum	0	0
Maximum	8,454	17,770
Percentiles		
25	3.2	0.4
50	11.2	4.5
75	159.6	15.6

In the case of distances greater than 100 kilometres, the spatial patterns in Toronto reflect the high number of personal contacts located at regional, national, and international spatial scales. In contrast, although there is a much lower proportion of social contacts outside Concepción, some egos maintain relationships with alters that live very far, with maximum values that also reflect the high distances between Chile and other places of the world. As discussed before, these figures are also consequence of the much higher levels of migration within Canada and the higher proportion of immigrants in Toronto, with respect to Concepción.

In sum, the spatial patterns of personal networks present an interesting perspective of social activity-travel in cities, placing enriching traditional explanations of social trips, only based on population fields or probabilities (Moore and Brown 1970), incorporating a rich set of explanations both from the individuals' viewpoint. In fact, the high variability of the spatiality of personal networks reflects the high range of distances of social trips and communication, which people perform to maintain their social networks. As distance between people is still important for frequency of interaction (Mok et al. 2010), these results also remark the role of the transport system on enabling social trips at different spatial scales.

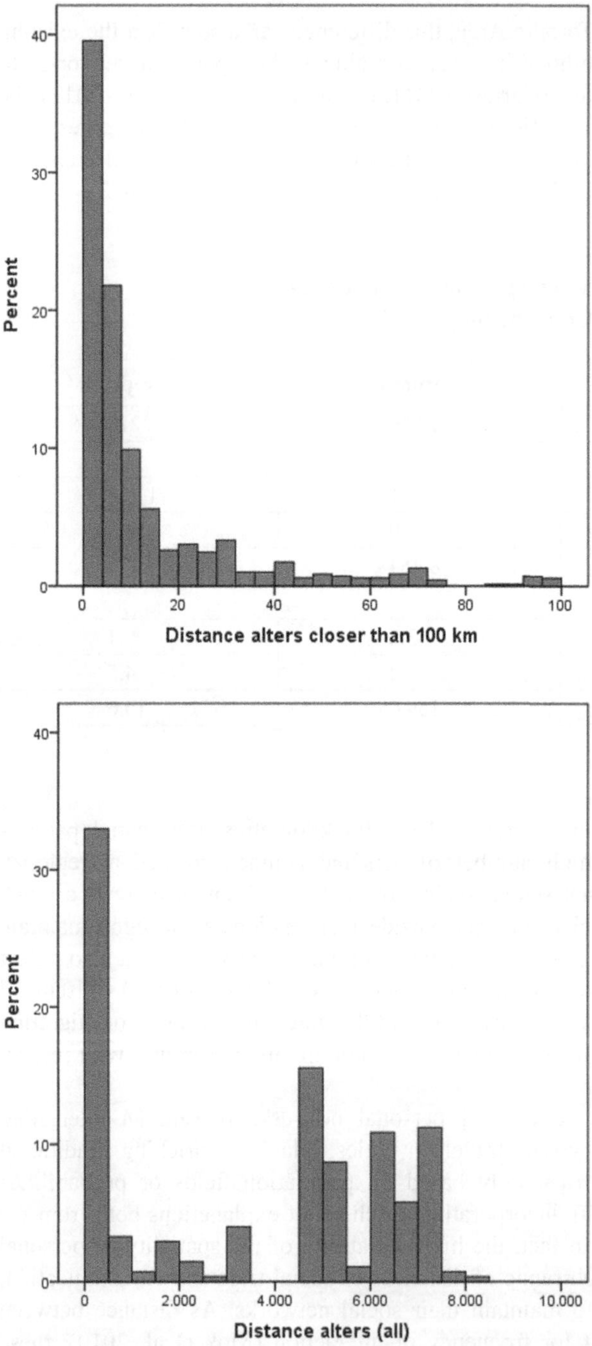

Figure 6.1a and b Distribution of alters in Toronto, according to distance

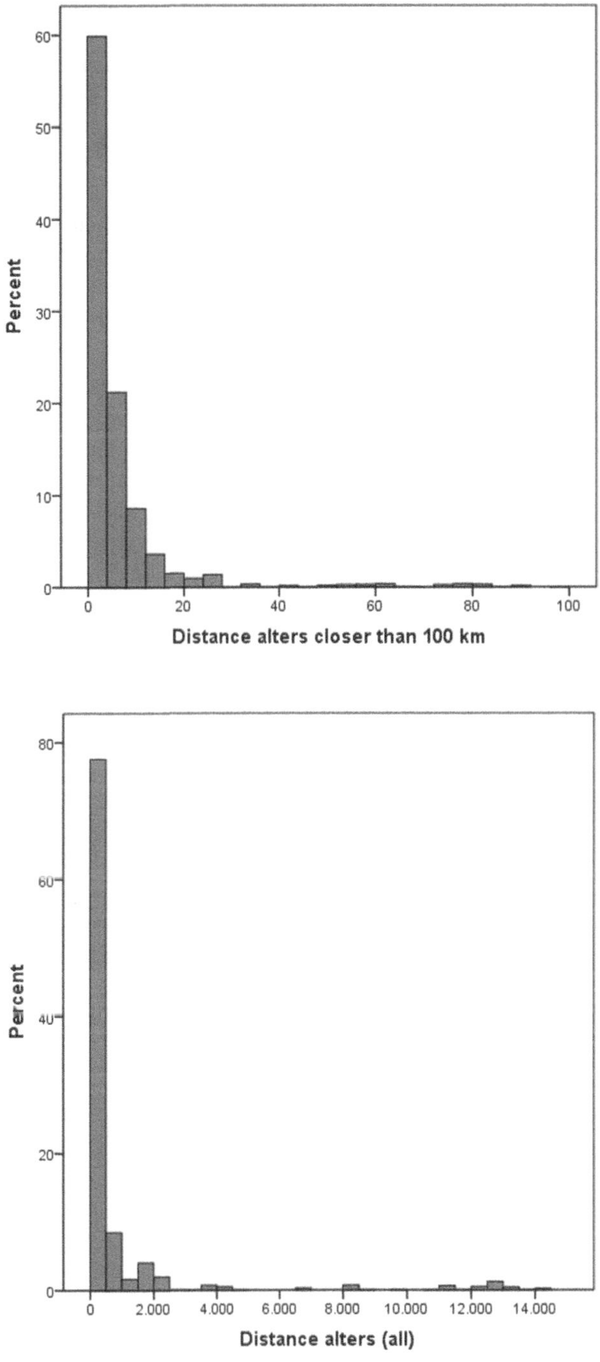

Figure 6.2a and b Distribution of alters in Concepción, according to distance

The previous results also underline the importance of the specific city and national context where social relations are embedded, which an aspect that transportation models and policies do not always take into account. In fact, a life course approach which could take into account previous residences and job, including international migration is key to understand social trip patterns spatially, as reflected by the previous results, and earlier discussions in the literature (e.g. Axhausen 2007). In addition, the ability to maintain personal networks in space has a very clear material nature, that is, it depends heavily on people's resources, including money and mobility tools.

The study of the spatiality of personal networks also links with discussions about the role of transportation systems on broader urban issues such as social exclusion and spatial segregation. In fact, as Marques (2012) shows in the case of Sao Paulo, Brazil, there are sharp differences on the spatial scale of personal networks among different income groups, which could reflect broader effects in terms of social isolation and social support. Obviously, from a transportation viewpoint, this reality also reflects the difference on the specific characteristics and needs of different income groups in the city. In order to briefly explore this issue, Table 6.4 presents the ego-alter distances, disaggregated by the four neighbourhoods that were surveyed in Concepción, which differ among them in terms of income and closeness to the CBD.

Table 6.4 Distances between ego and alters per network by neighbourhood type in Concepción

Distance	Neighbourhood type			
	Low income, close to CBD	*High income, close to CBD*	*Low income, far from CBD*	*High income, far from CBD*
Mean	65	449	70	213
Mode	0	455	453	453
Standard Deviation	505	1,996	219	809
Minimum	0	0	0	0
Maximum	11,255	17,770	2,718	12,708
Percentile 25	0.0	1.5	0.3	3.7
Percentile 50	0.3	5.8	2.0	7.0
Percentile 75	7.1	103.1	14.4	216.5

The results from Table 6.4 show the sharp differences between income groups, remarking how personal networks in low income groups tend to be, in average, much local than their high income counterparts. The mean distance and percentile 75 on both low income neighbourhoods (65 and 70 km for the median, and 7 and

14 for the percentile 75), reflect how most of the social contacts have a urban nature, which contrasts with high income neighbourhoods that not only have high mean distances (449 and 213 km), but also a urban scale much closer to the percentile 50 (6 and 7 km).

This spatial distribution of personal contacts reflects a number of processes that need to be taken into account when studying transportation systems. For example – as discussed before – mobility geography is an important long term process that shapes social relations over time. In the case of Concepción, migration processes are differentiated by income groups, since most of high income migrants have moved from further locations to the city than their low income counterparts, who present, in general, more local movements over time (Aroca and Rodríguez 2013). A second aspect that can be hypothesized with these results is the relevance of mobility tools and the contrast between car availability – mostly for high income groups – and the heavy reliance or 'captivity' on public transport – mostly for low income groups – that defines people's availability to encounter others. On that regard, the high locality on low income groups can be the outcome of a dual, reinforced process. In fact, local patterns of persona networks on low income groups can be affected by long term immigration and residential location processes, influencing the location of family and friends. At the same time, the lack of good access to transport (e.g. extreme reliance on public transport, transit fares as a barrier or closeness to services) can also constrain the access to people in space, reinforcing this trend over time.

Another interesting result from Table 6.4, which remarks the utility of network approaches to broaden the traditional perspectives to study transport systems, is the difference on the spatial distribution of personal networks according to closeness to Concepción's CBD. As the centre of this city is still the main source of jobs, leisure facilities, and residential supply, this spatial closeness is linked with the potential social encounters with others and, particularly, with the possibility of walking – and not relying on motorized transport – for social trip purposes. As the results show, people living at far distances from the CBD have, in average, longer distances with their alters, suggesting their need of travelling more to have social encounters with their personal networks, who are slightly more spread in space.

Time and Personal Networks: Frequencies of Interaction

All previous analyses and hypotheses regarding the usefulness of studying the spatial patterns of personal networks to understand transport systems do not present the whole picture if we do not consider the time component, in terms of the frequency of interaction between social contacts. Furthermore, maintaining personal networks require an appropriate regularity of face to face interaction in time (Wellman et al. 2006), as well as the complement of information and telecommunication technologies (Carrasco 2011; Rainie and Wellman 2012).

From a transportation viewpoint, studying frequencies of interaction also relates to social trip generation, that is, how many trips are performed for social purposes.

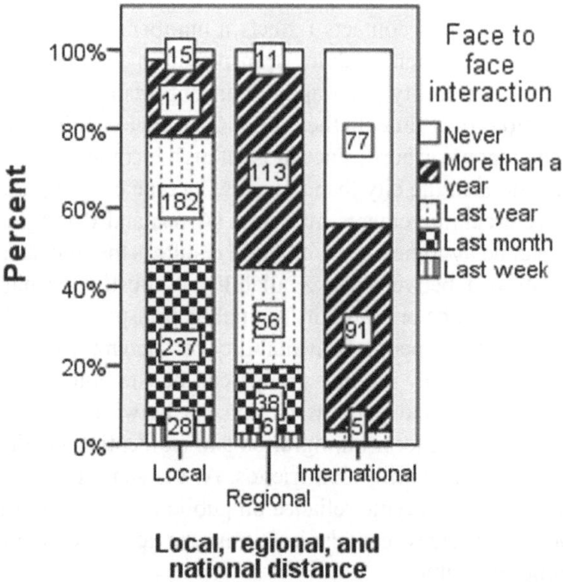

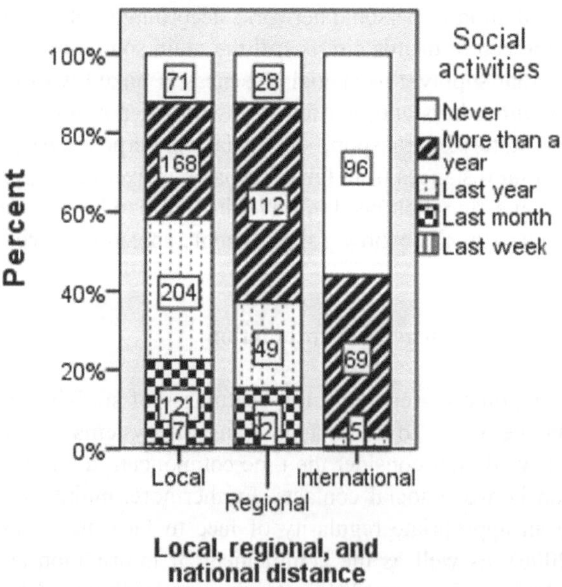

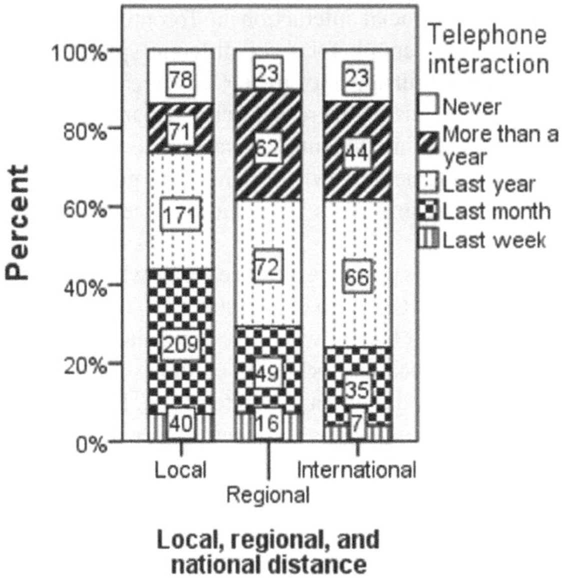

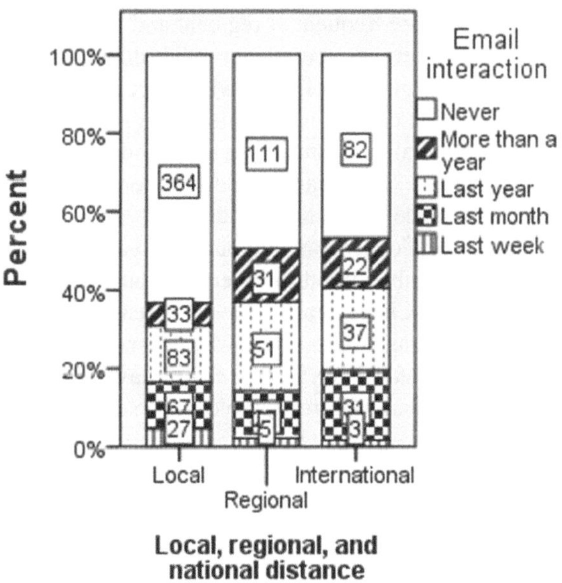

Figure 6.3a, b, c and d Frequency of interaction face-to-face, socializing, emailing and by telephone in Toronto, according to spatial scale

Figure 6.3 presents frequencies of social interaction in Toronto by face-to-face, socializing, by telephone, and by email, for three different spatial scales: local (at the metropolitan), regional (outside the greater city area, but at the national level). Face to face and socializing were explicitly separated since a main interest is differentiating between instrumental interactions (such as those existing in workplaces) and social interactions (visiting, hosting, going to pubs and restaurants); also capturing the circumstances when face to face interactions are different from socializing.

Both face to face and social interactions present a distance decay, which has been documented with great detail elsewhere (Carrasco et al. 2008b; Mok et al. 2010), showing that distance is still an important barrier when for social interaction. The integration of information and communication technologies shows the complex interplay between spatial scale, media, and frequency of interaction. For example, face-to-face interactions are proportionally more frequent at local scales, and yearly face-to-face encounters are proportionally very similar between regional and international alters. Socializing, on the other hand, show similar trends, although with a higher share of those alters who never interact (i.e., never perform social activities) with their egos. Monthly interactions by telephone occur at a similar share regardless of the spatial scale, but daily and weekly phone interaction has a higher share with local alters than with regional or international. Finally, email interaction seems to be overall more frequent at regional and international scales, compared with the local scale. However, these latter trends do not seem to be too conclusive due to the overall high number of alters who never interact with egos using these media, especially in the case of instant messaging. Overall, the relationship between face to face and socializing, on the one hand, and information and communication technologies, on the other hand, follows a complementary pattern, that is, technologies do not replace interaction in person.

Similar graphs can be produced for Concepción; although the values will be obviously different, the previously described trends are very similar (see Figure 6.4). The main potential difference in the Chilean case is the previously mentioned role of mobility tools availability. In fact, as discussed elsewhere, cars help people to have more social interaction (Farber and Páez 2009), and the availability of telephone and Internet connection can have a differential role on personal network maintenance (Boase 2008; Carrasco 2011).

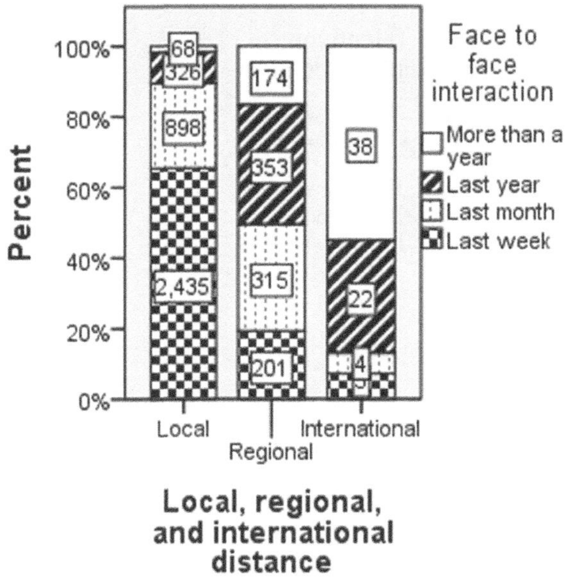
Face to face interaction
More than a year
Last year
Last month
Last week
Percent
100%
80%
60%
40%
20%
0%
68
326
898
2,435
174
353
315
201
38
22
4
Local
Regional
International
Local, regional, and international distance

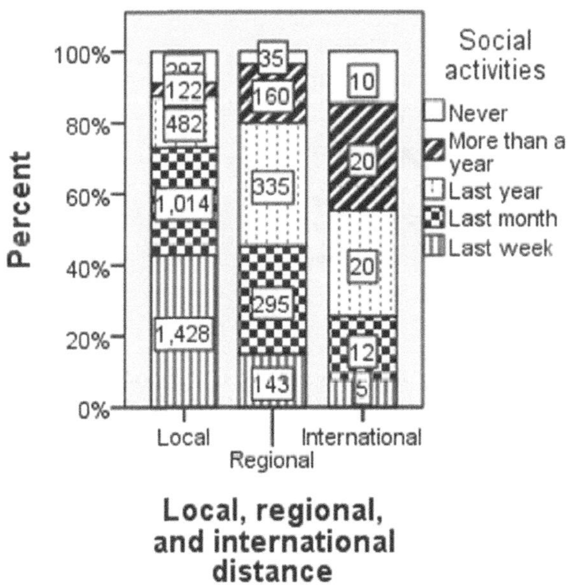
Social activities
Never
More than a year
Last year
Last month
Last week
Percent
100%
80%
60%
40%
20%
0%
122
482
1,014
1,428
35
160
335
295
143
10
20
20
12
5
Local
Regional
International
Local, regional, and international distance

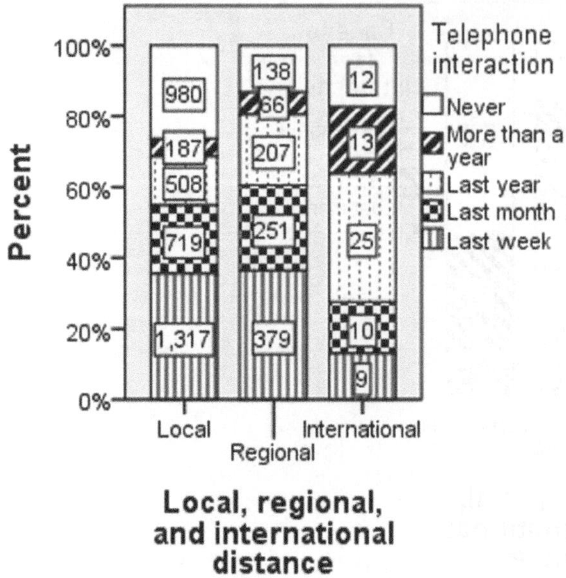

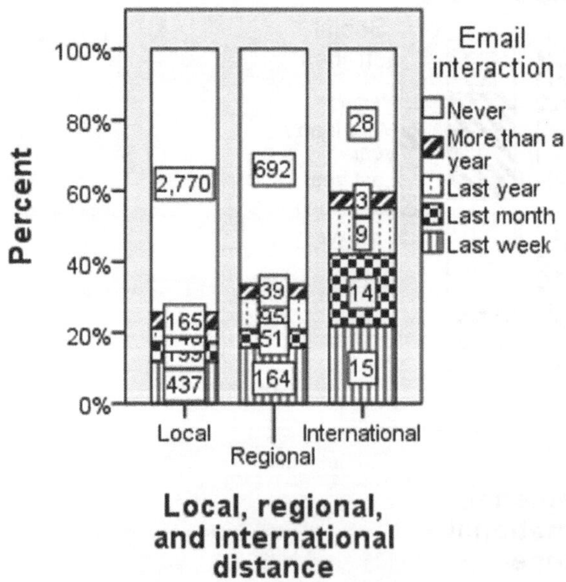

Figure 6.4 a, b, c and d Frequency of interaction face-to-face, socializing, emailing and by telephone in Concepción, according to spatial scale

Conclusions

Descriptive statistics and discussions presented on this section, suggest how understanding personal networks can help to understand social activity-travel in fundamental ways that go beyond the traditional individual, economic lens. In fact, understanding the people's social environment through their personal network characteristics, inspire hypotheses about the characteristics of social travel demand, by linking size, density, and sub-grouping to the need of maintaining different social foci, and thus having more social activity in time and space. With that perspective, an explicit consideration of ego-alter home distances shows the local based nature of personal networks on the one hand, but also the higher variability, depending on the city, income, and neighbourhood. This variance reflects the potential relevance of mobility biography processes, individual's mobility tools, and other contextual aspects, often neglected on travel behaviour research. This role of distance on social interaction, is also even clearer when frequency of interaction is also inserted on the analysis, reflecting not only that distance still matters for social trips, but that information and communication technologies play a complementary role on these processes.

References

Aroca, P. and Rodríguez, J. (2013) Population distribution and internal migration issues in LAC. In: J.R. Cuadrado-Roura and P. Aroca (eds) *Regional Problems and Policies in Latin America*, Advances in Spatial Science. Berlin: Springer Verlag, pp. 341–58.

Axhausen, K.W. (2007) Activity spaces, biographies, social networks and their welfare gains and externalities: Some hypotheses and empirical results. *Mobilities* 2: 15–36.

Boase, J. (2008) Personal networks and the personal communication system: Using multiple media to connect. *Information, Communication & Society* 11: 490–508.

Carrasco, J.A. (2011) Personal network maintenance, face to face interaction, and distance: Studying the role of ICT availability and use. *Transportation Research Record: Journal of the Transportation Research Board* 2231: 120–28.

Carrasco, J.A. and Cid-Aguayo, B. (2012) Network capital, social networks and travel: An empirical illustration from Concepcion, Chile. *Environment and Planning Part A* 44: 1066–84.

Carrasco, J.A., Bustos, C. and Cid-Aguayo, B. (2013) Affective personal networks versus daily contacts: Analyzing different name generators in a social activity-travel behaviour context. In: J. Zmud, M. Lee-Gosselin, M. Munizaga and J.A. Carrasco (eds) *Transport Survey Methods: Best Practice for Decision Making*. London: Emerald Press.

Carrasco, J.A., Hogan, B., Wellman, B. and Miller, E.J. (2008a) Collecting social network data to study social activity-travel behaviour: An egocentred approach. *Environment and Planning B* 35: 961–80.

Carrasco, J.A., Miller, E.J. and Wellman, B. (2008b) How far and with whom do people socialize? Empirical evidence about distance between social network members. *Transportation Research Record: Journal of the Transportation Research Board* 2076: 114–22.

Farber, S., Páez, A. (2009) My car, my friends, and me: A preliminary analysis of automobility and social activity participation. *Journal of Transport Geography* 17: 216–25.

Feld, S. (1981) The focused organization of social ties. *American Journal of Sociology* 86: 1015–35.

Freeman, L. (1979) Centrality in social networks: Conceptual clarification. *Social Networks* 1: 215–39.

Hogan, B., Carrasco, J.A. and Wellman, B. (2007) Visualizing personal networks: Working with participant-aided sociograms. *Field Methods* 19: 116–44.

Lyons, G. and Kenyon, S. (2003) Social Participation, Personal Travel and Internet Use. Proceedings of 10th International Conference on Travel Behaviour Research, Lucerne, 10–15 August 2003.

Marques, E. (2012) *Opportunities and Deprivation in the Global South: Poverty, Segregation and Social Networks in São Paulo*. Farnham: Ashgate.

Mok, D., Wellman, B. and Carrasco, J. (2010) Does distance matter in the age of the Internet? *Urban Studies* 47: 2747–83.

Moore, E. and Brown, L. (1970) Urban acquaintance fields: An evaluation of a spatial model. *Environment and Planning* 2: 443–54.

Rainie, L., Wellman, B. (2012) *Networked: The New Social Operating System*. Cambridge, MA: MIT Press.

Wellman, B., Hogan, B., Berg, K., Boase, J., Carrasco, J.A., Coté, R., Kayahara, J., Kennedy, T. and Tran, P. (2006) Connected lives: The Project. In: P. Purcell (ed.) *The Networked Neighbourhood*. Berlin: Springer.

Chapter 7
Dynamic Social Networks and Travel

Fariya Sharmeen, Theo Arentze and Harry J.P. Timmermans

Introduction

A social network is a representation of an individual's social connectedness. Fields as distinct as health, psychology and marketing are involved in social network research. A recent promising addition is the field of travel behaviour. In travel behaviour research, traditional factors of interests are facets of travel choice (such as frequency and transport mode) and ownership of mobility resources (such as car, bicycle, discount cards, period travel passes). Gradually, the concern shifted from trip-based to activity-based approaches to model travel properly as a derived demand from the activities that people conduct in space and time. The attention also shifted from individuals to households. Consequently, joint activity scheduling, task allocation, and resource allocation were incorporated in the choice models (Borgers et al. 2002; Ettema et al. 2004; Schwanen et al. 2007; Zhang et al. 2005). However, joint activities do not only involve household members, but may also include members of a person's social network. Often, we negotiate with our friends and family about where to go for holidays, who should host the New Year party or what movie to go to this weekend. Each individual is part of social networks and individual behaviour will be influenced by peer groups. Spatial behaviour analysis is incomplete without an understanding of this social dimension. To better understand people's activity-travel patterns, we need to understand how people select and organize their social contacts, adding a whole new dimension of transport behaviour modelling research. However, there is another distinct, and often ignored, feature of personal social networks: it is dynamic. It changes with time and with life course. In this Chapter, we emphasize the need to explain social networks and corresponding activity behaviour in a dynamic perspective. We conclude the chapter with a discussion of constraints and benefits of incorporating these dynamics and suggest directions of future research.

Towards Dynamics

The relationship between social networks and travel has been the subject of an increasing number of studies in recent years (Carrasco and Miller 2005; Hackney and Axhausen 2006; Han et al. 2011; Páez et al. 2008; Silvis et al. 2006; van den Berg et al. 2009). Most of these studies have examined the cross-sectional relationships between characteristics of social networks and facets of physical and virtual travel. However, the contemporary challenge in activity-travel behaviour analysis is to move beyond single-day approaches to dynamic activity-based models, focusing on adaptation to demographic, social and policy changes. The next challenge is to incorporate social network dynamics as the evolving nature of social networks and corresponding activity travel agendas have not been addressed in any detail in transportation research and beyond.

Why do we need to move towards dynamics of social networks? Social networks induce, alter, and even constrain travel. This effect is not static; it evolves continuously. We do not have the same social networks all our life. Our circle of friends, neighbours and contacts keeps changing as we age and life cycle events (e.g. marriage) happen. To understand how long-term behaviour and preferences change, it is imperative to understand these patterns in a dynamic setting. Assuming social networks are constant over time would fundamentally be erroneous in the first place and could result in flawed predictions of travel demand models. Studies show that social networks may induce adaptation in activity and travel choices. It is therefore imaginable, that changes in social networks can induce changes in activity-travel choices as well. For instance, Arentze and Timmermans (2008) conducted numerical simulations and argued that participation in social networks may lead to adaptation of aspirations and diffusion of knowledge, which in turn may trigger changes in activity-travel choice behaviour. Han et al. (2011) elaborated and extended this approach and demonstrated that the exchange of information and the formation of network-specific aspiration levels can influence choice set formation and destination choice behaviour. Páez and Scott (2004) and Hackney and Axhausen (2006) also used simulation approaches to demonstrate the possible effect of social networks on activity choices.

There are a number of ways through which social networks may affect transportation choices (Figure 7.1). First, social networks provide travel companions for social and leisure trips in particular. Individuals need to discuss and find agreement with their companions in decisions on the when, where and how aspects of activity and travel. Secondly, social networks are an important source of information. Not only for short term travel choices but also for long terms decisions, such as finding a new home or job (Connerly 1985, 1986; Dawkins 2006; Granovetter 1995), social networks play a key role. Thirdly, networks form an individual's primary source of social and emotional support, ranging from painting the house to child/elderly care (Lai 2001; Wellman and Frank 2001). Fourthly, peers have an influence on evaluations of choice options. Whether for novelty or approval seeking individuals tend to put value to how their alters or the society in general perceive their choice options (Abou-Zeid and Ben-Akiva 2011; Abou-Zeid et al. 2013; Schmöcker et al. 2013). ICT and social media support and shape the social interactions involved. With the growing influence of ICT in our lives, getting in touch or planning an activity is faster and easier (Mokhtarian 2002; Senbil and Kitamura 2003; Dijst 2009; Tillema et al. 2010; van den Berg et al. 2010). Despite the growing influence of ICT, face-to-face contacts and hence the geographic distribution of social contacts is still an important factor in the formation and maintenance of social relationships. If we or members of our social network move, we travel more or less, further or closer, weekdays or weekends to socialize with them. When social networks change it potentially brings changes in any or all of the ways, eventually changing individual's activity and travel patterns.

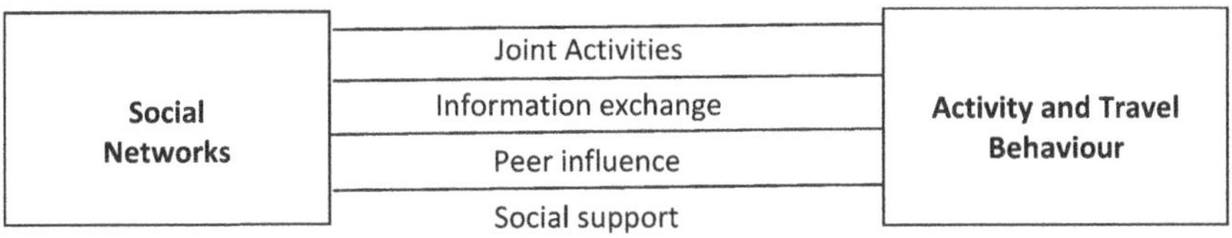

Figure 7.1 Interactions between social networks and activity-travel behaviour

The importance of dynamic personal networks has been long acknowledged in the field of sociology and demographic research (Hummon 2000; Snijders et al. 2010; Watts 1999), but it is not readily evident how such research can be elaborated to fit the agenda of transportation research. Here, we summarize contemporary efforts and in the next section we illustrate an empirical example of modelling social network dynamics in transportation research.

Approach and Major Findings

Sharmeen et al. (2010) focused on long term dynamics of social networks in the transportation literature. The motivation of this work stems from the simple notion that social networks are not invariant over time or over the life course of an individual. The key feature of the research is, therefore, long term dynamics. Life cycle events are assumed to be triggers of these dynamics.

Conceptual Framework

A central concept of the research is that dynamics of social networks are related to life cycle events, and have an effect on the activity and travel behaviour dynamics. The conceptual framework, therefore, is concerned with the following three domains:

1. The life-cycle domain
2. The social network domain, and
3. The activity and travel domain

It states that all these three domains are dynamic and interrelated. Change in any one domain can have an effect on the others. For example, when a person gets married (a life-cycle event), (s)he gets new in-laws and friends, as a result of which his/her social network may change. Since the person has a partner now the household maintenance jobs (such as grocery shopping, picking up/dropping off dry cleaning, etc.) can be distributed between the partners. Associated with the new social network, there can be new social visits and recreational activities. These have a direct impact on the activity and travel needs of the person. The effect can also be direct, for instance when the activity-travel schedule changes due to a job change. Let us consider the three domains as three parallel lines running along with each other as presented in Figure 7.2.

Certainly causality may also be in reverse direction. For instance, marriage and conjugal relationships and new jobs can be mediated through social networks, changing activity and travel behaviour may necessitate relocation, etc. However, the scope of the research is limited to investigate, firstly whether social networks change with life cycle events and secondly if they do, whether this has any effect on activity and travel needs. Investigating the causalities of life cycle events is beyond the scope of the research. Given this focus, the data collection is aimed at changes in social networks. The purpose of Figure 7.2 is to convey that these three domains are interrelated. This is an area not put forward before in the literature. Although all three domains can influence each other, the research focuses on social network changes occurring after a particular life-cycle event.

The framework is developed to capture these notions. Life-cycle changes are assumed to have an effect on social networks and activity-travel needs. Events may bring in changes in individual's time budget or location status or both. For instance, residential relocation includes a change of address and a modified geographical distance with all existing social network members. For some ties, these changes are substantial and will cause a change in interaction frequency. As a result, the tie may eventually disappear from the ego's active social network. Similarly, events may add new ties in the network. Therefore, effects of an event on the number of lost ties and number of new ties are expected.

Events may also have an effect on time budget. This may induce stress in the overall equilibrium bringing the system out of balance. Individuals then need to reschedule their activity-travel agenda. Hence, the effects of events on activity-travel needs are also taken into consideration.

In terms of social network dynamics, the framework assumes that there is a threshold of social network size. An ego cannot keep on adding ties indefinitely. Since the time the ego needs to maintain those ties is limited, at some point some ties should fade away. With this assumption, the model tests the effect of new ties on lost ties. Furthermore, the interdependencies between social network dynamics and activity-travel needs are investigated. Details of the framework and investigation are reported in Sharmeen et al. (2014a).

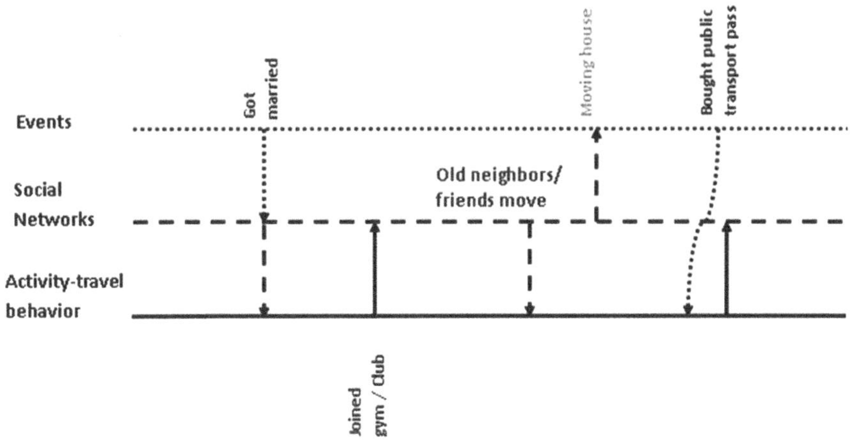

Figure 7.2 Conceptual framework representing three domains of interrelated dynamics

Source: Sharmeen et al. (2010).

Data Collection

Perhaps the most difficult aspect of investigating social network dynamics is to obtain the data. Ideally, panel data would be needed, but such surveys are very demanding. Therefore, in this research project we used an event-based retrospective survey to serve the purpose (see Chapter 2; also see Ohnmacht 2004). Respondents were pre-screened based on life cycle events. They were asked to choose one recent event from the event list provided at the beginning of the survey. Recent events were defined as events that took place within the past two years. If multiple events were experienced, then the respondent was asked to choose the most recent one to answer the questions. If they qualified, respondents were forwarded a set of questions related to the event. To capture such effects in the data, we carefully chose the period of two years so that respondents would have a new settled activity-travel pattern comparable to their pre-event agenda. Events were selected based on a literature review (Bidart and Degenne 2005). Presumably, all five events have an impact on social network and social interaction patterns because time and money budgets and social needs are changed.

During sampling, we aimed to have an equal representation of each type of life cycle event. However, after correcting for non-response and eliminating incomplete questionnaires, we obtained between 18% and 21% responses for each life cycle event. The average response rate was 56%. Table 7.1 presents a summary.

The web-based and paper-based surveys were administered in September 2011 in the Netherlands. The survey was divided into four parts seeking information about socio-demographics of the respondent, the present social network, changes and new contacts (if any) in the personal social network and in activity-travel behaviour in response to the life-cycle event in question. The first part collected socio-demographic details of the respondent including an estimate of the size of their present social network according to type of relationships. The second part asked details about their close ties. Close ties were defined as those individuals with whom important information is shared, personal problems are discussed, help during emergency or daily necessities is asked and with whom regular contact exist. Respondents could mention up to 25 ties. Details include socio-demographics of the alter, geographical distance, frequency of interaction, relationship strength (on a five point scale) and duration (in years).

Table 7.1 Summary of life cycle events

Type of life cycle event	# participants	# Completed questionnaire	# Non-response	% completed questionnaire	% of total completed questionnaire
Relocation	231	135	96	58.4	19.2
New job	244	150	94	61.4	21.3
Change in civil status	194	134	60	69	19.1
Starting university	226	131	96	57.9	18.6
Children start school	351	153	198	43.6	21.8
Total	1,246	703	543	56.4	100

In the third part, respondents were asked whether any change in the social network occurred as a result of the event. If yes, they were forwarded to a table where they had to list existing ties that involved a change and new ties that were formed. Furthermore, for each listed tie they had to fill out the type of change (geographical distance, frequency of contact per mode both before and after the event), the socio-demographics of the alter, information about the tie (strength, length known) for each of the alter where a change occurred. They also reported new ties and lost ties here. If there were no changes with a tie, they were not asked these details. The fourth part of the survey asked about changes in activity-travel schedules before and after the event. In this survey, new ties were not included, because information about distance change was not available for new ties. It only concerns those ties that existed pre-event and did not changed post-event.

Aiming at broadening our view from static to dynamic social networks, this survey has systematically considered changes in social networks, social interaction and activity-travel patterns with life cycle events. The purpose of the data collection was to test and validate the conceptual framework, i.e. the existence and nature of the interdependencies between the identified domains by means of empirical data. First, we investigated the effects of life cycle events on social networks, social interactions (including effects of ICT) and activity-travel behaviour. Then, in an integrated framework, we explored the dynamic interrelationships between the domains of social networks and activity travel profiles, triggered by life cycle events.

Summary of Findings

We identified three interrelated domains of life cycle events, social network and activity-travel behaviour and argue that understanding dynamic repertoires of these three interconnected domains is crucial to predict long-term travel demand. The general approach was to test and validate the conceptual framework (Figure 7.2), i.e. the existence and nature of the interdependencies with empirical data. First, we investigated the effects of life cycle events on social networks, social interactions (including effects of ICT) and activity-travel behaviour. Then, in an integrated framework, we explored the dynamic interrelationships between the domains of social networks and activity travel profiles, triggered by life cycle events.

Social interactions may involve a decoupling between persons in space and time to some extent (Carrasco et al. 2008), since one does not need to be in the same place and time to communicate with each other owing to the growth of electronic means of communication (e.g. via e-mail, mobile telephone, online social media). To some extent because distance still matters in the maintenance of social networks (Dijst 2009; Mok et al. 2010). Our first set of analyses investigates face-to-face social interaction frequency after changes in geographical distance and contexts (accessibility, density) due to life cycle events. Owing to these changes social interaction mode and frequencies may change among two actors. At first, we investigated how geographical features and changes in geographical distance affect social interaction frequency. Additionally, we tested if the history of interaction frequency might have an effect. We developed an ordered logit model of face-to-face social interaction frequency incorporating homophily, accessibility and path dependence attributes (Sharmeen et al. 2014b).

The findings (of the study reported in detail in Sharmeen et al. 2014b) imply that urban densities influence social interaction/activity frequency differently. Effects of distance to transport infrastructures also vary. Moreover, number of shopping (non-grocery) facilities within one kilometre of ego's residence influence social interaction/activity frequency positively. However, the quantities of café/restaurants and attraction facilities in the same radius (within 1 km radius around the postcode) have a negative effect on social interaction/activity frequency. The main take away from these findings is that local accessibility indicators affect social interaction frequencies. Therefore local geographical attributes should be taken into due consideration to better explain the social interactions and travel. These factors can contribute in creating sociable cities and cohesive communities, as argued in recent studies (Farber and Li 2013; Farber et al. 2013).

In terms of the long-term dynamics in face-to-face social interactions, one of the major findings of the aforementioned study is that social interaction/activity frequency is path dependent. Individuals tend to maintain their social ties with similar frequencies as they did before. However, when life cycle events have taken place, social preferences, needs and aspirations may change or get constrained. Previous studies support the finding that individuals keep the frequency of social interaction with family and stronger ties on a similar level (Sharmeen et al. 2012). This implies that it is unlikely that ties with whom one meets daily would disappear from the social activity agenda.

However, changes in geographical distance triggered by life cycle events have significant influence on social interaction frequency. Frequency of contact decreases with ties that become distant after an event. Analysis of the interaction effects of distance changes between actors and type of life cycle events reveal some interesting results. For instance, social interaction frequencies with ties who are geographically close after a life cycle event are influenced by the type of event. If it is residential relocation, the meeting frequency increases whereas the opposite is the case for events related to a change in work/study hours. The findings add to the literature in sociology about distance decay (also see Chapter 5) in social relationships (Dennis 1977).

Secondly, we investigated the dynamics of interaction between physical and electronic modes of communication with alters (Sharmeen et al. 2013b). The relationship between face-to-face and ICT contact frequency was found to be of a complementary nature in previous research. However, in most of the studies, it was found or assumed that ICT use has an effect on face-to-face communication. In Sharmeen et al (2013c) we allowed for the endogeneity effects between the modes of communication with a multilevel path analysis model (Hox and Kreft 1994; MacKinnon 2008; Mueller 1996). We found evidence that the effects could also be in the other direction. In line with contemporary researches, we found that ICT has a complimentary effect on face-to-face communication. Adding to it, we found that face-to-face communication has a substitution effect on ICT interaction frequency in a dynamic context.

Moreover, there were some important findings regarding the dynamics of social network due to life-cycle events. The most intuitive one is that the number of new ties has a negative effect on the number of old ties. As more and more new members are added to one's personal social network, the loss of existing ties becomes more probable. Next, the relationships between lifecycle events and dynamics of time allocation to activity and travel was investigated (Sharmeen et al. 2013a). Using path analysis, we analysed the effect of several lifecycle events on changes in activity and travel time allocation. Significant and susbtantial impacts of lifecycle events on time allocations was observed. Moreover, the effects are varied in direction and magnitude according to the type of event.

The findings confirm the three hypotheses regarding lifecycle events and activity travel dynamics. Events related to subsistence activities, such as new job and starting university, obviously affect time allocation to subsistence activities.

Events related to big household changes, such as residential relocation and children starting school, influence time allocation to maintenance and leisure activity-travel. Events related to changes in civil status, which potentially changes the social network size and composition, such as, getting married/separated/divorced/beginning co-habitation, influence the time allocation to maintenance and leisure activities. We also have found evidence that all events affect time allocation to leisure activities in varying intensity and direction. This is most likely because leisure activities are the most flexible and discretionary of all. However, this finding also acts as an indicator that lifecycle events are associated with social networks since they are key in most leisure activity planning. Any change in leisure activity time allocation potentially means change in the type and pattern of maintenance of social networks, which in turn affects their size and composition. Consequently, dynamics of social networks are associated with dynamics of activity travel behaviour.

Furthermore, the relationships between social network dynamics and activity travel needs in response to life cycle events was examined. The framework presented in Figure 7.2 was tested using a structural equation model (Sharmeen et al. 2014a). The results confirm the expectation that activity and travel dynamics are influenced by life cycle and social network dynamics. Moreover social network and activity travel dynamics are interdependent, i.e. a change in one leads to a change in the other. Furthermore, the study finds that travel needs are for the most part influenced by activity needs. Most socio-demographic and life cycle event variables have a direct impact on activity needs, which in turn generate travel needs.

These findings are consistent with the notion that activity generates travel. Travel does not have an impact on activity needs. We find this to be valid in a dynamic model as well. Furthermore, the findings related to social networks suggest that there is a optimum size for personal networks. The optimum may vary among individuals and perhaps also throughout life stages. A detailed study of this concept is needed and remains on the future research agenda.

Finally, we proposed a model to predict the formation of and changes in social ties among two actors in the population (Sharmeen et al. 2013c). The empirical analysis demonstrates how such a model can be estimated. Using the model and predictions, social networks for a population can be simulated and can be updated accordingly to accommodate changes of individual's and household's life cycle. The method and findings are relevant not only for long term travel behaviour analysis and prediction but also for human social behaviour in a more general sense.

The model derived from previous work (Arentze and Timmermans 2008; Arentze et al. 2012; Arentze et al. 2013) is based on fundamental theories and observations of social networks, such as homophily, reciprocity, transitivity and degree distribution (also see Chapter 4). To determine the social ties between two actors in the population it uses the random utility maximization assumption. The utility of a tie is measured by the degree of similarity between the actors and the geographical distance between them. A tie between two individuals is worthwhile only if the utility exceeds a threshold value for both persons involved. Thus the model predicts if a new tie emerges or not between two individuals. However, social ties and social networks are dynamic and may change especially in relation to life cycle events. Therefore, the model is extended to also predict behaviour in an adaptation stage in case a life cycle event occurred.

Findings suggest that the formation of new ties is influenced by homophily between the actors in the initial phase. In addition to that, geographical proximity influences the formation of a tie. The effects however are minimized in the adaptation phase, in the sense that sensitivity to homophily declines or in some cases a marginal inclination towards heterophily can be observed. However, for existing ties the effects of heterophily become stronger.

The model and predictions can be used to simulate social networks for a population wide distribution (Arentze et al. 2013). One can create a social network in the initial phase and then update it accordingly using the type of life cycle events. Further existing social ties could be updated using the history effect predictions.

The empirical analyses provided evidence that these domains are interconnected (Sharmeen et al. 2014a). Social network dynamics affect activity-travel needs and vice versa. The numbers of lost and new ties and time allocations to activity and travel vary depending on the types of life cycle events (Sharmeen et al. 2013a). Moreover, the numbers of new and lost ties are interdependent indicating the existence of an optimal personal social network size. In terms of social interaction, it was found that face to face social interactions are influenced by change in geographical distance to alters and accessibility and urban density of an ego's home location (Sharmeen et al. 2014b). Furthermore, it was found that ICT and face to face social interactions influence each other in the long term (Sharmeen et al. 2013b).

Summarizing the findings, we observed that social networks are dynamic and so are the activity-travel choices, needs and aspiration levels. Evidently, they influence each other. These long term dynamics and their interdependencies are essential for a comprehensive understanding of travel behaviour. Modelling with the assumption of static or independent nature of what are actually dynamic dimensions should lead to biases in prediction. Social policies and travel demand forecasting models are designed for the long run, making these incorporations relevant. Of course, we are far from completely grasping the interactions and time dynamics; nonetheless the study indicates the relevance of the dynamics for transportation research and travel demand forecasting models.

Limitations

Most of the study limitations relate to the data and assumptions related to the data collection. Some typical errors and biases are associated with retrospective surveys, which were inevitable in this case. However, studies have shown that retrospective data can be useful if carefully collected. When studying mobility issues, panel data contain a higher risk of inconsistency and sometimes discontinuity related to nonresponse and selective loss of mobile respondents is employed (Solga 2001). On the other hand, studies show that with proper design and execution respondents can recall events and socio-demographics with a higher degree of accuracy (Behrens and Mistro 2010; Berney and Blane 1997). In order to reduce errors, the survey was carefully designed and administrated with the aim of reducing respondent burden, focusing on one particular recent event and the changes associated with it to ensure respondents have a clear idea when answering the questions. Three rounds of pilot testing were organized and detailed feedback reports from respondents and academics were taken into account to improve the design at each step. Applying possible and pragmatic measures, it was attempted to ensure that the responses have better consistency and as few errors as possible. Yet, it is likely that there are still errors and memory biases particularly in the area of reporting durations before the events for less important activities.

There were some missing data related to lifecycle events (key events such as child birth and details of residential moves) in the analysis. This was primarily due to lack of data. We acknowledge that situational changes may have occurred during the after-event period that are not recorded and hence cannot be included in the analysis. The objective of the chapter (and data collection) is to capture the effects of a major life cycle event to the activity-travel scheduling. People need some time to absorb and incorporate the effect of the event in their activity-travel budget. To capture such effects in the data, we carefully chose the period of two years so that respondents would have a new settled activity-travel programme comparable to their pre-event agenda. The survey was based on the above assumptions.

We also assumed that life cycle events trigger long-term changes in social networks and activity travel needs. Although this assumption is based on literature review, there are other reasons that may bring in changes in social networks, which could not be incorporated in the study.

Social networks and interactions are a two-way street. In this research project, we only hear and include one side of the story. Similar life cycle events and changes may occur to the alters. Their home locations, urban density and accessibility features may also have an influence on social interactions and travel, which we could not account for. Neither do we have detailed information about social media use and their effects.

Challenges Remain

In recent years, researchers have addressed several aspects of the influence of social network on activity and travel planning, such as duration, frequency, start time, distance, mode choice, etc. The reported results are intuitive and demand due attention by researchers as well as policy makers and practitioners to incorporate the social context in activity and travel demand forecasting. However, it is a long way to completely understand the social dimensions of activity and travel. Therefore challenges remain. Here, we summarize the broader topics.

A first challenge is to incorporate social networks in application oriented travel demand forecasting models, so as to predict social and leisure travel demand with greater understanding. To see the big picture, we need to broaden our view from egocentric social networks (for an example see Chapter 3) to a population-wide projection (e.g. see Chapter 4). An agent based simulation environment is a potential platform to serve the purpose and should be utilized accordingly. Further, to simulate the potential of tie formation, the negotiation of a social meeting (mode choice, place, time etc.) has been simulated in transportation research using multi-agent simulation. However our knowledge is still limited. In this study, we have extended ways to predict population-wide social network dynamics, which needs further modification, for instance, incorporating geographical attributes, and sensitivity tests to assess the error and biases associated with replication.

Another predicament in social networks analysis is obtaining the necessary data. To get a good impression of the heterogeneity of activities, it is desirable to collect longitudinal diary data. Traditional travel diary surveys collect one or two days of travel data from participants. While cross-sectional travel diary surveys are useful in determining the overall average travel behaviour of the regional population, they do not capture repetitive patterns in social activities, for instance weekly routines of people. New GPS technologies and use of smartphones may provide a promising way of collecting longitudinal travel data without asking too much effort from respondents.

In addition, the study of the effect of ICT's on social activity-travel remains a topic for future research. Although this topic has been studied recently, the possibilities of ICT's are increasing rapidly. These changes will affect joint travel, for instance in arranging a joint trip (e.g. making reservations, buying tickets, checking routes, weather and travel conditions), making additional research necessary. With the growing influence of social media in individuals' social life, dynamics of social network has become important more than before. Possibly virtual social networks are more flexible, at least in terms of new ties, where social media plays an important role in information dissemination, publicity of social events, etc., having further impact in activity and travel needs. We could not incorporate the effects of social media in particular. This could be done by collecting information about virtual social networks, interaction and travel patterns. Further peer influence on travel choices could also be investigated by tracking information dissemination in social media. The possibilities of extension are therefore diverse and promising.

A further challenge is to link the a-spatial aspect of social network to the spatial one. There is a spatial facet attached to an individual's social network as far as travel and transportation is concerned. The distance and accessibility of the peers should matter in planning and maintaining social networks. In this chapter, we only had data from the ego, not from the alter, which is a very relevant and possible future extension. Snowball sampling method can be used to collect the relevant information (an example for such a study is provided in Chapter 4). Alternatively targeted stated adaptation questionnaires could be designed to understand the role of geography and infrastructure on social interaction to a greater depth.

Moreover, studies have found support to the assumption that personal social network tends to change with major life cycle events. New ties are added and some old ties fade away. They also suggest that there is a threshold of the size of personal social network. The threshold may vary among individuals and perhaps also throughout life stages. A detailed study of this concept is needed and remains in the future agenda.

Finally, due consideration should be attached to social network dynamics in choice set development. In travel behaviour modelling identifying choice sets is a challenge as individuals are not aware of a universal choice set when selecting an activity-travel profile. They cannot even always consider all of those they are aware of, constrained by their habitual preferences (path dependence) or resource availability. Information about attitude and personality may assist to understand these characteristics. Moreover, the choice sets are dynamic and major change is induced by the evolving social networks.

The analysis of social networks has a far reaching potential in understanding almost all aspects of human behaviour. These potentials have been discussed for some time. A recently growing field of exploration is activity and travel behaviour. The literature of social networks and transportation literatures is nonetheless still at an exploration stage. There have been some commendable works already. Further comprehension and integration of the local social context, social externalities and social dynamics to the travel behaviour models remain on the contemporary agenda.

Acknowledgements

The research leading to these results has received funding from the European Research Council under the European Community's Seventh Framework Programme (FP7/2007–2013) / ERC grant agreement n° 230517 (U4IA project). The views and opinions expressed in this publication represent those of the authors only. The ERC and European Community are not liable for any use that may be made of the information in this publication.

References

Abou-Zeid, M. and M. Ben-Akiva (2011) The effect of social comparisons on commute well-being, *Transportation Research Part A: Policy and Practice*, 45(4): 345–61

Abou-Zeid, M., J.-D. Schmöcker, P.F. Belgiawan and S. Fujii (2013) Mass effects and mobility decisions, *Transportation Letters*, 5(3):115–30

Arentze, T. and H. Timmermans (2008) Social networks, social interactions, and activity-travel behavior: A framework for microsimulation, *Environment and Planning B: Planning and Design*, 35(6): 1012–27

Arentze, T., P. van den Berg and H. Timmermans (2012) Modeling social networks in geographic space: Approach and empirical application, *Environment and Planning A*, 44(5):1101–20

Arentze, T.A., M. Kowald and K.W. Axhausen (2013) An agent-based random-utility-maximization model to generate social networks with transitivity in geographic space, *Social Networks*, 35(3): 451–9

Behrens, R. and R.D. Mistro (2010) Shocking habits: methodological issues in analyzing changing personal travel behavior over time, *International Journal of Sustainable Transportation*, 4(5): 253–71

Berney, L.R. and D.B. Blane (1997) Collecting retrospective data: Accuracy of recall after 50 years judged against historical records, *Social Science & Medicine*, 45(10): 1519–25

Bidart, C. and A. Degenne (2005) Introduction: The dynamics of personal networks, *Social Networks*, 27(4): 283–7

Borgers, A., F. Hofman and H. Timmermans (2002) Conditional choice modelling of time allocation among spouses in transport settings, *European Journal of Transport and Infrastructure Research*, 2(1): 5–17

Carrasco, J.A., B. Hogan, B. Wellman and E.J. Miller (2008) Agency in social activity interactions: The role of social networks in time and space, *Tijdschrift voor economische en sociale geografie*, 99(5): 562–83

Carrasco, J.A. and E.J. Miller (2005) Socializing with people and not places: Modelling social activities explicitly incorporation social networks. *Proceedings of the 9th International Conference on Computers in Urban Planning and Urban Management (CUPUM)*, London

Connerly, C.E. (1985) The community question: An extension of Wellman and Leighton, *Urban Affairs Review*, 20(4): 537–56

Connerly, C.E. (1986) What should be done with the public housing program?, *Journal of the American Planning Association*, 52(2): 142–55

Dawkins, R. (2006) *The Selfish Gene*, Oxford: Oxford University Press

Dennis, R.J. (1977) Distance and social interaction in a Victorian city, *Journal of Historical Geography*, 3(3): 237–50

Dijst, M. (2009) ICT and social networks: Towards a situational perspective on the interaction between corporeal and connected presence, in R. Kitamura,

T. Yoshi and T Yamamoto (eds) *The Expanding Sphere of Travel Behaviour Research*, New York: Emerald Publishing, pp. 45–76

Ettema, D., T. Schwanen and H. Timmermans (2004) The effect of locational factors on task and time allocation in households. *Proceedings of the 83rd annual meeting of the Transportation Research Board January 11–15, 2004*, Washington, DC

Farber, S. and X. Li (2013) Urban sprawl and social interaction potential: An empirical analysis of large metropolitan regions in the United States, *Journal of Transport Geography*, 31: 267–77

Farber, S., T. Neutens, H.J. Miller and X. Li (2013) The social interaction potential of metropolitan regions: A time-geographic measurement approach using joint accessibility, *Annals of the Association of American Geographers*, 103(3): 483–504

Granovetter, M. (1995) *Getting a Job: A study of Contacts and Careers*, Chicago: University of Chicago Press

Hackney, J.K. and K.W. Axhausen (2006) An agent model of social network and travel behavior interdependence. *Proceedings of the 11th International Conference on Travel behaviour Research (IATBR)*, Kyoto

Han, Q., T. Arentze, H. Timmermans, D. Janssens and G. Wets (2011) The effects of social networks on choice set dynamics: Results of numerical simulations using an agent-based approach, *Transportation Research Part A: Policy and Practice*, 45(4): 310–22

Hox, J.J. and I.G. Kreft (1994) Multilevel analysis methods, *Sociological Methods & Research*, 22(3): 283–99

Hummon, N.P. (2000) Utility and dynamic social networks, *Social Networks*, 22(3): 221–49

Lai, G. (2001) Social support networks in urban Shanghai, *Social Networks*, 23(1): 73–85

MacKinnon, D. (2008) *Introduction to Statistical Mediation Analysis*, New York: Routledge

Mok, D., B. Wellman and J. Carrasco (2010) Does distance matter in the age of the Internet?, *Urban Studies*, 47(13). 2747–83

Mueller, R.O. (1996) *Basic Principles of Structural Equation Modeling: An Introduction to LISREL and EQS*, New York: Springer

Ohnmacht, T. (2004) Soziale Netze und persönliche Mobilität: Grundlagen für eine empirische Erhebung, *Arbeitsbericht Verkehrs-und Raumplanung*, 250

Páez, A., D.M. Scott and E. Volz (2008) A discrete-choice approach to modeling social influence on individual decision making, *Environment and Planning B: Planning and Design*, 35(6): 1055–69

Schmöcker, J.-D., T. Hatori and D. Watling (2013) Dynamic process model of mass effects on travel demand, *Transportation*, 41(2): 279–304

Schwanen, T., D. Ettema and H. Timmermans (2007): If you pick up the children, I'll do the groceries: spatial differences in between-partner interactions in out-of-home household activities, *Environment and Planning A*, 39(11) 2754–73

Sharmeen, F., T. Arentze and H. Timmermans (2013a): Incorporating time dynamics in activity travel behavior model: A path analysis of changes in activity and travel time allocation in response to life-cycle events, *Transportation Research Record: Journal of the Transportation Research Board*, 2382(1): 54–62

Sharmeen, F., T. Arentze and H. Timmermans (2013b) A multilevel path analysis of social network dynamics and the mutual interdependencies between face-to-face and ict modes of social interaction in the context of life-cycle events, in M.J. Roorda and E.J. Miller (eds) *Travel Behaviour Research: Current Foundations, Future Prospects*, Toronto: Lulu Press, pp. 411–32

Sharmeen, F., T. Arentze and H. Timmermans (2013c) Population-wide social network dynamics with life-cycle events: Modeling approach and first analyses. Paper presented at the *International Workshop: Frontiers in Transportation–Social Interactions*, Herrsching

Sharmeen, F., T. Arentze and H. Timmermans (2014a) An analysis of the dynamics of activity and travel needs in response to social network evolution and life-cycle events: A structural equation model, *Transportation Research Part A: Policy and Practice*, 59: 159–71

Sharmeen, F., T. Arentze and H.J.P. Timmermans (2012) Out of sight, out of mind? Investigating changes in mode and frequency of social interaction due to lifecycle events. *Proceedings of the 13th International Conference on Travel behaviour Research (IATBR)*, Toronto

Sharmeen, F., T.A. Arentze and H.J.P. Timmermans (2014b) Dynamics of face-to-face social interaction frequency: role of accessibility, urbanization, changes in geographical distance and path dependence, *Journal of Transport Geography*, 34(1): 211–20

Silvis, J., D. Niemeier and R. D'Souza (2006) Social networks and travel behavior: Report from an integrated travel diary. *Proceedings of the 11th International Conference on Travel Behaviour Reserach*, Kyoto

Snijders, T.A., G.G. Van de Bunt and C.E. Steglich (2010) Introduction to stochastic actor-based models for network dynamics, *Social Networks*, 32(1): 44–60

Solga, H. (2001) Longitudinal surveys and the study of occupational mobility: Panel and retrospective design in comparison, *Quality & Quantity*, 35(3): 291–309

van den Berg, P., T.A. Arentze and H.J.P. Timmermans (2009) Size and composition of ego-centered social networks and their effect on geographic distance and contact frequency, *Transportation Research Record: Journal of the Transportation Research Board*, 2135: 1–9

Watts, D.J. (1999) Networks, dynamics, and the small-world phenomenon 1, *American Journal of Sociology*, 105(2): 493–527

Wellman, B. and K.A. Frank (2001) Network capital in a multilevel world: Getting support from personal communities, in N. Lin, K.S. Cook and R.S. Burt (eds) *Social Capital: Theory and Research*, Chicago: Aldine de Gruyter, pp. 233–73

Zhang, J., H.J. Timmermans and A. Borgers (2005) A model of household task allocation and time use, *Transportation Research Part B: Methodological*, 39(1): 81–95

Index

Page numbers in italic indicate figures and bold indicates tables.